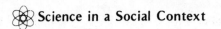 Science in a Social Context

Scientific Images and their Social Uses
An Introduction to the Concept of Scientism

Iain Cameron

and

David Edge
Science Studies Unit, University of Edinburgh

Butterworths
LONDON · BOSTON
Sydney · Wellington · Durban · Toronto

The Butterworth Group

United Kingdom	Butterworth & Co (Publishers) Ltd
London	88 Kingsway, WC2B 6AB
Australia	Butterworths Pty Ltd
Sydney	586 Pacific Highway, Chatswood, NSW 2067
	Also at Melbourne, Brisbane, Adelaide and Perth
Canada	Butterworth & Co (Canada) Ltd
Toronto	2265 Midland Avenue,
	Scarborough, Ontario, M1P 4S1
New Zealand	Butterworths of New Zealand Ltd
Wellington	T & W Young Building,
	77–85 Customhouse Quay, 1, CPO Box 472
South Africa	Butterworth & Co (South Africa) (Pty) Ltd
Durban	152–154 Gale Street
USA	Butterworth (Publishers) Inc
Boston	19 Cummings Park, Woburn, Mass. 01801

All rights reserved. No part of this publication may be reproduced or transmitted in any form or by any means, including photocopying and recording, without the written permission of the copyright holder, application for which should be addressed to the Publishers. Such written permission must also be obtained before any part of this publication is stored in a retrieval system of any nature.

This book is sold subject to the Standard Conditions of Sale of Net Books and may not be re-sold in the UK below the net price given by the Publishers in their current price list.

© SISCON 1979
First published 1979
ISBN 0 408 71309 7

British Library Cataloguing in Publication Data

Cameron, Iain
   Scientific images and their social uses.
   1. Science — Social aspects
   I. Title   II. Edge, David   III. Science in a Social Context (Project)
   301.24'3   Q175.5   78-41272

ISBN 0-408-71309-7

Typeset by Butterworth Litho Preparation Department
Printed in England by Billing and Sons Ltd,
Guildford and London

Introduction

In modern industrial societies, science plays a central and respected role. Scientific knowledge is taken to be the final word on, and is often called on to arbitrate over, 'what is really the case'. To 'be scientific' is to earn admiration: it is to claim (and often to gain) credibility and superiority over unscientific alternatives — while to 'be unscientific' is to be indefensibly woolly, vapid, old-fashioned, inefficient and generally unworthy of serious consideration. The community of scientists (which produces, legitimates and disseminates scientific knowledge) enjoys high status, substantial rewards and a position of power. One result of this central position of the scientific institution is that our culture is now suffused with notions about, and images of, science and scientists. These notions and images are common property. We all have a conception of what 'being scientific' implies, and when someone uses the phrase we summon up shared associations and meanings. Similarly, aspects of scientific knowledge itself have diffused into commonsense: words such as 'evolutionary', 'atomic', 'nuclear', 'feedback' and 'ecological' are in everyday use. Such concepts and images, we might say, are *resources*, freely available for all of us to deploy.[1]* When we think it appropriate, we draw on these resources, assembling them to suit our purposes. And when we do so, we are said to be practicing *scientism;* the discourse which arises is then *scientistic*.

Definitions are never wholly satisfactory, but we can say that scientism is present *where people draw on widely shared images and notions about the scientific community and its beliefs and practices in order to add weight to arguments which they are advancing, or to practices which they are promoting, or to values and policies whose adoption they are advocating.*

The concept of scientism implies an *attitude to science:* those who use scientistic language acknowledge and respect *the authority of the scientific community,* and wish to capitalize on that authority, in order to make their discourse more persuasive. In so doing, they reinforce and consolidate that authority. You might either applaud or deplore such activities, or consider them well or badly done; but it is important to appreciate that you can examine the way scientism operates without taking sides. The concept, and analysis to which it gives rise, is intended to be essentially *descriptive,* to characterize 'what is going on'. Instances of scientism can be recognized and discussed in much the same way as we can trace the development of folk music or popular art. This book is intended as an introduction to both the concept and

---

* Notes to the Introduction can be found on pp. 6—8.

the analysis: it is designed to get you started on this way of thinking about the relationship between scientific knowledge and society, and to indicate the kind of understanding and illumination which emerges from this approach.

However, the term is not always (indeed, is seldom) used in this purely descriptive or analytic way. Those who espy scientism in others are normally using the concept as *a form of criticism*. Often, scientism is *defined* as the *illegitimate* use of images drawn from science to add *inappropriate* weight to arguments in which such implicit appeals to scientific authority *should have no place*, as, for instance, in van Hayek's definition of scientism as 'the *slavish imitation* of the method and language of science'.[2] Here, *the use of the concept* itself implies *an attitude to science:* namely, that science 'has its proper place' and that the authority of the scientific community should be *carefully contained*. To adopt this *evaluative* usage of the notion of scientism is to *object* that the scientific community is being imbued with *too much* authority (perhaps, even, being treated as a source of *sacredness*). And so we find the common complaint that too much weight and value are attached to the concepts, methods and results of the natural sciences: that, for instance, in trying to make the exercise of political judgment 'more scientific' by the use of quantitative techniques purporting to maximize human and social benefits, an aspect of science is being *inappropriately* carried over into a domain in which its application is *illicit*.[3] Thus, at the simplest level, the belief that science can eliminate all human suffering and misery, and hence by itself promote human happiness, would be dubbed scientistic by one who thought that science was incapable of satisfying man's religious, esthetic or moral needs. Alternatively, it is claimed that scientistic thought relies on a mistaken view of the nature of scientific method and practice. The common element in all these formulations is the charge that an *excessive respect* for the success, prestige and authority of science results in a *dangerous misconception* of its scope and validity.

However, this evaluative use of the concept of scientism should not blind us to the process to which it refers. Images of and from science *are* available, as cultural resources, to add substance to discourse, and people *do* use them in this way. One very common example is in the rhetoric of ethics: in this manifestation, criticism of scientism is particularly passionate, and the label 'scientistic' especially abusive. Many authors have attempted to use features of science to justify, endorse and promulgate a moral code or a set of ethical ideals and values. Their attempts can, broadly speaking, be divided into two categories. Some draw on widely-shared beliefs about the *practice of science* and the values which appear to underlie the (successful) development of the professional scientific community; others appeal to the set of facts, laws and theories (i.e. the beliefs) of that community and attempt to ground a moral code in *a scientific description of the world*. In Chapters 1 and 2 we will examine examples of these two sorts of argumentation, and raise questions about them — questions both about their *success*, as logical arguments, and about their *status*, as kinds of writing. These examples will convey something of the flavour or *nature* of scientism, in the sense in which we have defined it on page 3.

In Chapter 3, we will discuss material which sets scientistic discourse of these (and related) kinds in a social context and asks *why people are led to make use of scientific resources in this way*. The suggested answer, in outline, is: *to support, consolidate or advance their own social positions*. People are predisposed to make use of the available cultural resources to rationalize and justify beliefs and attitudes that they already (at least in principle) possess, and that are, essentially, a function of the social and political system of which they are a part. In modern society, science contributes its share to the available cultural resources — a share which can, perhaps, be over-emphasized (because of the present high social status of the scientific community), but which consists of materials of *essentially the same kind* as the other available resources. All these resources can be *manipulated*. We suggest that this analysis throws a fresh light on the relationship between science and society; it offers fruitful ideas for the exploration of contemporary problems in this area and of the rhetoric which surrounds them.

This book is intended to facilitate exploration of the topic by both students *and* teachers — hopefully as a co-operative venture. The Notes at the end of each chapter are designed to include sufficient leads to follow up to obviate any need for a separate Teacher's Guide. It is, of course, a necessarily brief and incomplete introduction; it is worth listing some of the issues which it neglects and briefly referring to some objections raised to our original draft.

1. We have paid no direct attention to what is often held to be the main contemporary manifestation of scientism — namely, the attempt to develop quantitative techniques of social forecasting and planning under such titles as 'policy sciences', 'systems analysis', 'technology assessment' and 'futurology'. However, there is now a fairly extensive literature which offers an analysis of these topics.[4] Nor have we dealt with criticisms of the infiltration of scientific classifications and methods into the practice of medicine and psychiatry[5], or such historical topics as Taylorism and the rise of scientific management[6]. The list of omissions could no doubt be extended indefinitely. Readers are encouraged to develop our approach in their own exploration of these, and other similar, controversies.
2. It has been pointed out to us that we have tended to use historical materials as if their interpretation were not a matter of contention among historians. In fact, our historical examples are *not* always

unproblematic, and we have tried to incorporate references in the Notes which argue for different interpretations, where these are known to us.[7] However, since our aim is to introduce the elements of a *kind* of analysis, criticism of the details of any particular example is not of crucial importance to the general form of the argument.

3. This book should, ideally, be set in a broad context of sociological analysis, and treat issues such as those raised by Marcuse and Habermas in a systematic and rigorous way. (Marcuse and Habermas are, of course, notoriously difficult to understand: their interpretation is at least as problematic as that of some of the materials we have selected.) Again, we have attempted to refer to these wider issues where it seemed appropriate to do so, but to do more is beyond our present scope. If the book tempts readers to venture more widely, its job will have been done.[8]

4. Scientism can be seen as an attempt to 'colonize' territory where scientific language, techniques, approaches, models and metaphors (unproblematically established elsewhere) have previously been thought inapplicable. But it has been argued that *science itself* also advances in this kind of way.[9] To those engaged in developing econometrics, for instance, the task is that of 'making economics more scientific', and their efforts are widely held to be legitimate, however scientistic they may seem to their critics; the extension into biological disciplines of techniques pioneered by physicists has raised similar controversies (and also won Nobel Prizes). Although we do not discuss the point, it is worth considering whether the same social and cultural mechanisms may not operate both *within* and *outside* the scientific community: the appropriateness of the new deployment of established scientific resources is then a matter of (differing) judgment by the several parties affected or involved.[10]

5. It has been put to us that, in stressing the descriptive and analytical (and value-free?) use of the concept of scientism, we have ourselves adopted a scientistic stance in writing this book. If by this it is meant that we have attempted to achieve some measure of traditional academic detachment, then we are not ashamed of this accusation.[11] But readers can form their own judgments on this matter.

Notes

1. Barry Barnes is a leading exponent of this notion, which derives from Levi-Strauss' metaphor of the *bricoleur*. See: Barnes, B. (1974). *Scientific Knowledge and Sociological Theory.* London: Routledge and Kegan Paul.

2. Von Hayek, F.A. (1942). 'Scientism and the Study of Society.' *Economica* **9**, 269 (our italics). The reader would do well to collect examples of such definitions, to classify them as descriptive or evaluative, and to decide (after completing the book) whether they are satisfactory!
3. Note here a parallel with the use of the Marxist notion of 'Ideology'. However purely descriptive this term *can* be, and however useful in social analysis, there is no doubt that most people *use* it *evaluatively*. To call someone's thought 'ideological' is to claim that it is in *error* — indeed, that it is *dangerously misleading*. For a discussion, see Barnes, Note 1, esp. Ch. 6, 'Science and Ideology'.
4. See, for instance:
    Cross, N., et al. (Eds) (1974). *Man-Made Futures: Readings in Society, Technology and Design*. London: Hutchinson Educational and the Open University Press.
    Hoos, Ida R. (1972). *Systems Analysis in Public Policy: A Critique*. Berkeley: University of California Press.
    Martin, Brian (1978). 'The selective usefulness of game theory.' *Social Studies of Science* **8**, 85–110.
    Tribe, Laurence H. (1972). 'Policy science: analysis or ideology?' *Philosophy and Public Affairs* **2**, 69–110.
    Wynne, Brian (1975). 'The rhetoric of consensus politics: a critical review of TA.' *Research Policy* **4**, 108–169.
5. See, for instance:
    Illich, Ivan (1975). *Medical Nemesis: The Expropriation of Health*. London: Calder and Boyars (revised and reprinted as *Limits to Medicine* by Marian Boyars, 1976, and republished in Pelican Books, 1977).
    Pappworth, M. H. (1967). *Human Guinea Pigs*. London: Routledge and Kegan Paul (also published by Penguin Books, 1969).
    Szasz, Thomas S. (1971). *The Manufacture of Madness*. London: Routledge and Kegan Paul (also published by Paladin, 1973).
6. See, for instance:
    Haber, Samuel (1964). *Efficiency and Uplift: Scientific Management in the Progressive Era 1890–1920*. Chicago and London: The University of Chicago Press (Midway Reprint paperback, 1973).
    Hales, Mike (1974). 'Management science and "The Second Industrial Revolution".' *Radical Science Journal* No. 1, 5–28.
    Maier, Charles S. (1970). 'Between Taylorism and Technocracy: European ideologies and the vision of industrial productivity in the 1920s.' *Journal of Contemporary History* **5**, 27–61.
7. For a useful review and bibliography of recent work in the social history of science, see: MacLeod, Roy (1977). 'Changing perspectives in the social history of science.' In *Science, Technology*

and Society: a Cross-Disciplinary Perspective. Ed. by Ina Spiegel-Rösing and Derek de Solla Price, pp. 149–195. London and Beverly Hills: Sage Publications

8. For excerpts from Marcuse and Habermas, see: Barnes, B. (Ed.) (1972). *Sociology of Science*, pp. 331–375. Harmondsworth, Middx.: Penguin Modern Sociology Readings. See also: Marcuse, H. (1964). *One-Dimensional Man*. London: Routledge and Kegan Paul; and Habermas, J. (1971). *Toward a Rational Society*. London: Heinemann Educational. Several SISCON Units are also relevant, and relate the present book to a wider context: Brian Wynne, *Sociology of Science* (3 Units and Teacher's Guide); Iain Cameron, *Metaphor in Science and Society;* Ron Johnston, *Science and Rationality* (2 Units); and George M. Schurr, *Science and Ethics*. Another very useful reader is: Teich, Albert H. (Ed.) (1977). *Technology and Man's Future*. New York: St. Martin's Press; among other things this contains a reprint of John McDermott's stimulating essay, 'Technology: The opiate of the intellectuals', originally published in *The New York Review of Books* (31 July 1969). See also: Young, Robert M. (1973). 'The human limits of nature.' In *The Limits of Human Nature*, pp. 235–274. Ed. by J. Benthall. London: Allen Lane; and Barnes, Barry and Shapin, Steven (1977). 'Where is the edge of objectivity?' (essay review of Mary Douglas, *Implicit Meanings*), *Br. J. hist. Sci.* **X**, 61–66.

9. See: Mulkay, Michael (1974). 'Conceptual displacement and migration in science: a prefatory paper.' *Science Studies* **4**, 205–234; and the discussion of metaphor in Barnes, Note 1.

10. The categories of 'scientific' and 'scientistic' can then both be seen as socially grounded and evaluatively deployed: the distinction between them (if any) is a matter of dispute, with a very broad no-man's-land of debatable instances. The scientistic use of scientific images in poetry no doubt produces many mediocre poems and some that even scientists can see are bad; but it has also produced some poetry which is widely accepted to be good. See: Bush, Douglas (1950). *Science and English Poetry: A Historical Sketch 1590–1950*. New York: Oxford University Press (also in paperback 1967); and Meadows, A. J. (1969). *The High Firmament: A Survey of Astronomy in English Literature*. Leicester: Leicester University Press. The criterion, of course, is whether or not it is good *poetry* in the eyes of those whose judgment in such matters is normally respected.

11. See also David Bloor's *apologia* on this point, in: Bloor, D. (1976). *Knowledge and Social Imagery*, p. 144. London: Routledge and Kegan Paul.

A Note on Readings

The bulk of the text of this book consists of commentary on a number of selected Readings. Two of these Readings are reprinted as Appendices, but these by themselves are insufficient to achieve the book's aims; convenient access to further reading material is necessary. These additional Readings can be categorized as follows.

1. *Essential:* without the opportunity to read these with care, the student has little or no chance of grasping the main points.
2. *Strongly recommended:* these consolidate and elaborate the main points, and access to them will greatly enhance the basic learning.
3. *Recommended:* these allow the student to develop his/her grasp of the major points.
4. *Background:* these provide students and teachers with a wide range of 'follow-up' material.

Unless access to the majority of material in categories (1) and (2) is freely available, it would be unwise to attempt to use this book. It is also recommended that at least a reasonable proportion of categories (3) and (4) can be easily consulted.

The additional Readings are as follows.

CHAPTER 1

Essential

Merton, Robert K. 'The institutional imperatives of science.' In *Sociology of Science*, pp. 65–79. Ed. by B. Barnes (1972). Harmondsworth, Middx.: Penguin Modern Sociology Readings.

Strongly recommended

Barnes, B. and Dolby, R. G. A. (1970). 'The scientific ethos: a deviant viewpoint.' *Archive of European Sociology* **11**, 3–25.

Recommended

Ellis, N. D. (1969). 'The occupation of science.' *Technology and Society* **5**, 33–41. Abridged in Barnes (Ed.), *Sociology of Science*, see above, pp. 188–205.

Mulkay, Michael (1969). 'Some aspects of cultural growth in the natural sciences.' *Social Research* 36, 22–52. Abridged in Barnes (Ed.), *Sociology of Science*, see above, pp. 126–141.

Mulkay, Michael (1976). 'Norms and ideology in science.' *Social Science Information* 15, 637–656

Salomon, J. J. (1971). 'The *Internationale* of science.' *Science Studies* 1, 23–42.

Storer, Norman (1966). *The Social System of Science.* New York: Holt, Rinehart and Winston.

Background

As listed in the Notes for Chapter 1.

CHAPTER 2

Essential

Colwell, T. B., Jr. (1970). 'Some implications of the ecological revolution for the construction of values.' In *Human Values and Natural Science*, pp. 245–258. Ed. by E. Laszlo and J. B. Wilbur. New York: Gordon and Breach.

Flew, A. G. N. (1967). *Evolutionary Ethics.* London: Macmillan Papermac.

Huxley, T. H. 'Society modified according to human standards.' In *Darwin and Darwinism*, pp. 209–215. Ed. by Harold Y. Vanderpool (1973). Lexington, Mass.: D. G. Heath.

Spencer, Herbert. 'Society conditioned by evolution.' In *Darwin and Darwinism*, pp. 109–208. Ed. by Harold Y. Vanderpool (1973). Lexington, Mass.: D. G. Heath.

Strongly recommended

Douglas, Mary (1970). 'Environments at risk.' *Times lit. Suppl.*, 4 December, 1419–1421 (reprinted in: *Ecology: the Shaping Enquiry* (1972). Ed. by J. Benthall. London: Longman; and in *Implicit Meanings*, pp. 230–248. London: Routledge and Kegan Paul).

Quinton, Anthony (1965). 'Ethics and the theory of evolution.' In *Biology and Personality*, pp. 107–131. Ed. by I. T. Ramsey. Oxford: Blackwell.

Toulmin, S. E. (1970). 'Contemporary scientific mythology.' In *Metaphysical Beliefs*, pp. 3–71. Ed. by A. MacIntyre. London: SCM Press.

Recommended

Barbour, Ian G. (1966). *Issues in Science and Religion*, pp. 93–96, 408–414. London: SCM Press.
Fay, Charles (1970). 'Ethical naturalism and biocultural evolution.' In *Human Values and Natural Science*, pp. 159–167. Ed. by E. Lazlo and J. B. Wilbur. New York: Gordon and Breach.
Nash, Roderick (Ed.) (1968). *The American Environment: Readings in the History of Conservation*. Reading, Mass.: Addison-Wesley.
Needham, Joseph (1943). 'Integrative levels.' In *Time: The Refreshing River*. London: Allen and Unwin.
Raphael, D. D. (1958). 'Darwinism and ethics.' In *A Century of Darwinism*, pp. 334–359. Ed. by S. A. Barnett. London: Heinemann.
Waddington, C. H. (Ed.) (1942). 'The relations between science and ethics.' In *Science and Ethics*. London: Allen and Unwin.

Background

As listed in the text and Notes for Chapter 2.

CHAPTER 3

Essential

King, M. D. (1968). 'Science and the professional dilemma.' In *Penguin Social Science Survey 1968*, pp. 34–73. Ed. by J. Gould. Harmondsworth, Middx.: Penguin Books.
Rosenberg, Charles E. (1966). 'Science and American social thought.' In *Science and Society in the United States*, pp. 135–162. Ed. by D. Van Tassel and M. G. Hall. Homewood, Ill.: Dorsey (revised and reprinted in Rosenberg, C. (1976). *No Other Gods: On Science and Social Thought in America*, pp. 1–21. Baltimore, Md. and London: The Johns Hopkins University Press; and abridged under the title of 'Scientific theories and social thought' in Barnes (Ed.), *Sociology of Science*, see above, pp. 292–305).

Strongly recommended

Bronowski, J. (1971). 'The disestablishment of science.' *Encounter*, July, 9–16, 96 (reprinted in *The Social Impact of Modern Biology*, pp. 233–247. Ed. by W. Fuller. London: BSSRS and Routledge and Kegan Paul).

Recommended

Berman, Morris (1977). *Social Change and Scientific Organization: The Royal Institution 1799–1844.* London: Heinemann.

Hahn, Roger (1971). *The Anatomy of a Scientific Institution.* Berkeley, Calif. and London: The University of California Press.

Rosenberg, C. (1974). 'The bitter fruit: heredity, disease and social thought in nineteenth-century America.' *Perspectives in American History* **VIII**, 189–235 (reprinted in Rosenberg, *No Other Gods*, see above, pp. 25–53).

Smith-Rosenberg, C. and Rosenberg, C. (1973). 'The female animal: medical and biological views of woman and her role in nineteenth-century America.' *Journal of American History* **60**, 332–356 (reprinted in Rosenberg, *No Other Gods*, see above, pp. 54–70).

Background

As listed in the text and Notes for Chapter 3.

In addition the following are both *recommended* as supplementary reading for the book as a whole:

Barnes, Barry (1974). *Scientific Knowledge and Sociological Theory.* London: Routledge and Kegan Paul.

Bloor, David (1976). *Knowledge and Social Imagery.* London: Routledge and Kegan Paul.

Chapter One
Ethics and the Practice of Science

In this chapter, our main task will be to examine closely two papers by authors who, in their separate ways, make the scientist claim that the values which they judge to be inherent in *scientific practice* are superior to those prevailing in other areas of human activity, and are therefore worthy of extension into a general code of ethics. We have selected these two papers from a wide range of similar writings because they seem to us to exemplify contrasting and relatively extreme forms of this kind of argumentation.

However, before we proceed to this task, we will first draw attention to an influential body of sociological literature which sets out to describe and codify *the norms and values of the scientific community*. Whatever the precise academic aims of its authors, this corpus achieves a kind of 'ideal picture' of that community, reflecting and cataloging the features of the behavior of scientists that constitute the community's 'public image'. These features are the cultural resources on which scientism can feed: they are the aspects of scientific practice which can be drawn on to support more general moral and ethical explorations.

We will go on to consider the two examples of attempts to propound a 'scientific ethic'. Although it is *not* our intention to offer any systematic exposé of such scientistic writings, we will raise questions as to the coherence and the logical basis (if any) of these two essays. And we will close this chapter with a brief look at more recent work (by sociologists and others) which criticizes the sociological tradition and questions the validity of the 'public image' itself.

The norms and values of the scientific community

Statements of the norms and values of scientists stem from the work of the American sociologist Robert K. Merton, especially as elaborated by later commentators — notably Bernard Barber and N. K. Storer.[1]*
These authors are intent on *describing* what they take to be the actual state of affairs within the scientific community: namely, that its activities are dominated and controlled by a 'scientific ethos', and that scientists normally act in accordance with that ethos. They do not intend their work as an *evaluation* of science, nor do they *advocate* the

* Note for Chapter 1 can be found on pp. 24–26.

widespread adoption of these norms and values. However, in propounding an Ideal Type they are confirming common beliefs about scientific practice.

Here read: Merton, R. K. 'The institutional imperatives of science.' In *Sociology of Science*, pp. 65–79. Ed. by B. Barnes (1972). Harmondsworth, Middx.: Penguin Modern Sociology Readings.

Merton defines four essential values which together comprise 'the ethos of modern science': Universalism, Communism, Disinterestedness and Organized Scepticism. Barber melds aspects of these into the over-riding ideal of Rationality: to him, to accept this value means to hold in high esteem...

> ... the critical approach to all phenomena of human existence in the attempt to reduce them to ever more consistent, orderly and generalized forms of understanding.

The key word here is 'critical'. This is contrasted with an 'unscientific' respect for tradition — the valuation of whatever is old and established simply because it *is* old and established. The scientist, on the other hand, believes in a 'spirit of free enquiry'. He has a duty to question all accepted beliefs, and to be particularly suspicious of any which are justified by custom or ritual.

Such an ideal of Rationality is often invoked by those who wish to celebrate the moral qualities of scientific practice. For example, I. I. Rabi, the distinguished American physicist, writes:

> To revere and trust the rational faculty of the mind, to allow no taboo to interfere in its operation, to have nothing immune from its examination, is a new value which has been introduced into the world. The progress of science has been the chief agent in demonstrating its importance and riveting it into the consciousness of mankind. This value does not yet have universal acceptance in this country or in any other country. But in spite of all obstacles, it will become one of the most treasured possessions of all mankind because we can no longer live without it.[2]

Note that Rabi considers scientific rationality to be universal in its scope and opposed to irrational authoritarian constraints on belief. He holds up this value as unique to science (which, even if it did not 'introduce' it, has certainly *established* it), and takes its universal extension to be so evidently desirable as to be inevitable.

Equally popular with scientistic theorists are the Mertonian norms of Universalism and Communism (or, in Barber's subtle transformation, Communalism). These are drawn on in the portrayal of science as a source of egalitarian and democratic ideals, and of international amity. Often, this Universalism is underpinned by the proposition that 'the physical laws are everywhere the same'. By this belief, the scientist (so it is argued) is bound to subject all contributions to his discipline to the same impersonal standards of evaluation. The race, religion, political outlook, nationality and sex of his colleagues do not enter into his evaluation of their work — nor do such considerations restrict entry into the scientific community, which rests only on merit and professional competence openly and objectively demonstrated. By Communism (reinforced by Disinterestedness), each scientist is duty bound to share his own research findings with his fellow scientists, so that scientific knowledge (which belongs to the whole community, rather than to any individual, group or nation) can grow by the gradual accretion of contributions from many diverse sources. The international collectivity of science transcends national and ideological differences.

Scientistic rhetoric draws freely on such images. Science is 'the great common and universal expression of humanity in this century'[3] or 'the most complete model of humanity without frontiers'[4]; and scientists, by virtue of their acceptance of these values, emerge as ideal diplomats. As Isaac Asimov, the popular science writer, puts it:

> The nations of the world are divided in culture, in language, in religion, in tastes, in philosophy, in heritage — but wherever science exists at all, it is the same science: and scientists everywhere speak the same language and accept the same mode of thought. Is it not, then, as a class, the scientists to whom we must turn to find the leaders in the fight for world government?[5]

Often such claims are combined with the idea (derived out of Organized Scepticism and Disinterestedness) that the scrupulous mental hygiene and emotional detachment of the scientist gives him added advantages:

> The scientist, by his competence in predicting possible outcomes of certain actions, and by his critical approach to the problems confronting him, and to their various candidate solutions, is peculiarly fitted to lead the way towards *rational* solutions of society's problems.[6]

The Mertonian tradition identifies a formidable array of norms and values within the practice of science. Scientists are, apparently (and are, apparently, *believed to be*), egalitarian, tolerant, open-minded, predisposed to collaborate across intimidating social barriers, emotionally detached and supremely rational. Not only is their community a model of international co-operation, but also of internal political organization. If Disinterested Scepticism might tend to fragmentation, its Organization (via Communism and Universalism) maintains the community's coherence. C. H. Waddington, for instance, speaks of the supreme success with which the ethos of science has managed to resolve the contradiction between freedom and order; and J. Bronowski refers with admiration to the way in which the society of science preserves its stability, whilst encouraging dissent and freedom of thought.[7] Criticism and controversy are, apparently, essential to the progress of science; but commitment to the emerging universal truth transcends ensuing upheavals and ensures that the community continues as a unified whole.

Scientistic writers claim that such characterizations of the scientific community are of universal ethical significance, exemplifying values and virtues which we would do well to copy. And although the Mertonians do not make any such ethical claim, they do propose their picture as an Ideal Type which uniquely characterizes the scientific profession, describing its normal functioning and congruent with (if not actually the *cause* of) its widely-respected success.

Questions

Consider the following two passages written by practicing scientists. Which of the Mertonian norms are mentioned, and which of the two extracts is more scientific in the strong sense, defined in the Introduction?

[Scientists] do not make wild claims, they do not cheat, they do not try to persuade at any cost, they appeal neither to prejudice nor to authority, they are often frank about their ignorance, their disputes are fairly decorous, they do not confuse what is being argued with race, sex, or politics.[8]

The new world opened up by science cannot be run without mortal danger by the old sentimental political methods, greed, lust for power and domination. It can only be run by the spirit which has built science itself, and if we want to stay alive we must rebuild our political and social thinking and institutions from the ground up in the spirit of science, which is that of human solidarity and mutual respect.

We often hear it said that science has no moral content. This is wrong, and if the world is in trouble today this is just because it has run away with the results

> of science, leaving its spirit and morals behind. Morals
> are the simplest rules of human behaviour, what to do
> and what not to do, and science has one very definite
> advice to give: if you have a problem, meet it as such, as
> a problem, knowing that problems are like equations
> which cannot be solved with blows or bombs, not even
> by atomic bombs... To solve a problem, meet it with a
> cool head, with uncompromising intellectual honesty,
> unbiassed by greed, fear or hatred, collect data and try
> to find the best solution. If you have an adversary, look
> upon him with respect as your associate with whom,
> together, you have to find the best solution. If this
> spirit would prevail at the peace talks in Paris, peace in
> Vietnam could rapidly be achieved.[9]

'Scientific approach to ethics' (Rapoport)

We turn now to the first of our two samples of attempts to derive an ethical code from a consideration of the practice of science. These two papers have been selected as relatively short and clear representatives of the genre. Readers are invited to make their own collections of similar material and to check for themselves whether our samples can be regarded as typical.

Here read: Rapoport, Anatol (1957). 'Scientific approach to ethics.' *Science* 125, 796–799 (reprinted as Appendix 1). **In doing so, note carefully** *the form of Rapoport's argument, and the kind of evidence he cites to support it.*

Note the following features of Rapoport's argument.

1. It rests on the claim that science is *unique* in its realization of human values: that scientists have, as it were, stumbled upon the secret of how to conduct human affairs:

> [There] is something unique about the ethics of scientific practice which makes it a peculiarly suitable basis for a more general system.

In Rapoport's view, this ethical code is 'superior' and 'more viable' than any of the alternative professional ethics or traditional moralities. And this claimed uniqueness is supported by reference to the Universalism of science — to the image of the scientific community as a *successful* and *amicable* model of human society, in contrast with the marked 'lack of success' of traditional societies.

2. Rapoport locates this uniqueness as stemming from *beliefs about the nature of the 'pursuit of scientific truth'*, or what he calls the 'strategic principles' of science. He himself is clearly an *empiricist*. To him, science is built on the assumption that there exists a 'unitary objective truth', which can be arrived at by the application of 'rules of evidence'. Anyone can (in principle) master these rules, and all who employ them will inevitably be led to the same conclusion on the basis of a particular set of data. Rapoport apparently believes that it is *this assumption which leads scientists to value and pursue consensus*, and not the other way around. (Although note that he says, puzzlingly, that the desirability of 'converting' an opponent is 'really not included in the ethics of scientific practice', but is 'nevertheless carried over into scientific practice from other areas'.) His summary is worth quoting:

> [The] ethical principles inherent in scientific practice are: the conviction that there exists objective truth; that there exist rules of evidence for discovering it; that, on the basis of this objective truth, unanimity is possible and desirable; and that unanimity must be achieved by independent arrivals at convictions — that is, by examination of evidence, not through coercion, personal argument, or appeal to authority.

Rapoport is clearly much impressed with what Barber has called the norm of Rationality, including Organized Scepticism. His summary makes clear that he interprets scientific activity in a thoroughly empiricist fashion.

3. The importance of this empiricism to his case for the uniqueness of scientific ethics is demonstrated in his argument for the Universalism of science by appeal to *common experiences* — 'irreducible', rock-bottom, unquestionable and universally shared:

> [The] irreducible answers to scientific questions are answers linked to those irreducible experiences which can, potentially, be shared by all mankind... To the scientist, the act of communication... is the basic ethical act... Science, then, is the only human activity which taps the really universal communality of human experience at the roots... The bid of scientific ethics for universal acceptance rests on the claim of science to be the first instance of a universal point of view about man's environment and, moreover, a point of view not imposed by coercion or even by power of persuasion or dramatic, personal example, but by its inherent, universal appeal to universal human experience, through being rooted in reliable knowledge.

Inasmuch as more modern philosophies of science have challenged the basic tenets of empiricism, e.g. by emphasizing the 'theory-laden' nature of scientific 'facts', or the role of 'paradigms' in directing scientific enquiry, they have weakened Rapoport's argument. However, Rapoport makes three other claims which are not so closely tied to his empiricism. The first is that scientific knowledge is unique in that 'the scientific system does not shrink from the shattering of its own foundations':

> The scientific outlook is *the only one capable of self-examination;* it is the only one that raises questions concerning its own assertions and methods of inquiry; the only one that is able to uncover provincial biases which govern our convictions and thus at least give us the opportunity to avoid such biases.

The second is that adoption of the scientific outlook is *irreversible:*

> There seems to be something universally satisfying about the scientific view. Once the vantage point of this view is attained, other views seem impoverished, provincial, naive.

The third is that the alternative to the adoption of his universal scientific ethic is *unstable:*

> The physicist can, for a time, use scientifically ethical means (that is, pursuit of objective truth) in the service of scientifically unethical goals (for example, imposition of coercion by war). But [this position is] unstable, and [is] doomed to extinction. It is impossible, in the long run, to hold provincial views while pursuing knowledge.

4. This leads to a final feature of Rapoport's position: it is *Utopian.* Rapoport's ethics consist of *one over-riding principle* — that of Non-Coercion — to which 'it is tempting to suppose that' all other moral and ethical rules must comply, and from which such rules all stem. Rapoport looks forward to a world dominated by this principle, and which is, apparently, without any *moral dilemmas,* i.e. no situations in which different and inconsistent ethical principles can *clash.* 'The pursuit of truth' will, by itself, induce a state of *utopian harmony.*

And this, in turn, reveals Rapoport's underlying notion of the *ideal society:* it is a society *without conflict,* united by common interests and suffused with a universal ethic. Clashes of interest are features of 'unstable states', which are 'doomed to extinction'.

This suggests a point for further reflection. Rapoport's essay is, as we have seen, scientistic, in that it deploys widely-available images of science; and Rapoport *organizes* these images in sympathy with a philosophy of science (Empiricism) and an implicit *political* model. Is this a common feature of scientism? Might not all philosophical theories of scientific method and, for that matter, sociological attempts to describe the scientific community, have such 'hidden' political bases?[10]

> **Questions**
>
> Which of the Mertonian norms does Rapoport draw upon?
>
> Do you think it likely that the scientific ethic sketched out by Rapoport would be at all useful in solving moral problems encountered in the course of our everyday lives?
>
> Re-read Rapoport's essay, noting all the different contexts in which he terms science as 'universal', in contrast with the 'provincial' or limited features of other bodies of belief. Does Rapoport construct a coherent argument for the 'uniqueness' of science? Are you convinced by it? Allowing, for the moment, that he *does* establish such a 'uniqueness', does he offer convincing reasons as to why the values inherent in such an *unique* activity should be *universally* adopted?
>
> Is Rapoport's characterization of science accurate? Is the evidence for scientific statements basic and available to everyone? Does science question its own foundations? If he is inaccurate in these respects, does his plea for a scientific ethic fail?
>
> Is there any historical evidence to support Rapoport's claim that the pursuit of truth in science is incompatible with the imposition of coercion by war?
>
> Do you believe that adoption of the scientific outlook is 'irreversible'?

'On the need for a scientific ethic' (Mesthene)

We now turn to a second writer, who shares Rapoport's concern that we should all learn ethical lessons from the successful practice of science, but who presents a sharply contrasting view of what those lessons might be.

Here read: Mesthene, Emmanuel (1947). 'On the need for a scientific ethic.' *Philosophy of Science* 14, 96–101 (reprinted as Appendix 2). As before, pay careful attention to *the form of his argument.*

Note the following features of Mesthene's paper.

1. Mesthene pays no attention to the 'sociology' of the scientific community — its coherence or its patterns of communication. Nor does he argue that science is *unique.* Instead, he draws a direct analogy between *the process of reaching a moral or ethical decision,* and *the process of scientific experimentation.*

> The fault of traditional morality lies in its failure to see that ethical principles are of the same *kind* as the laws and rules which guide scientists in other fields.

To practice a scientific ethic when faced with a moral dilemma, therefore, is to *experiment,* matching the facts of the situation to the available (and potentially incompatible) ethical principles *in essentially the same way as a scientist struggles to match his experimental data with the available theories.* This flexible approach allows us to *learn* in ethical matters, just as it allows the scientist to learn in *his* domain.

2. In Mesthene's view, scientific and moral principles are similar because they stand in the same sort of relationship to the concrete situations in which they are applied. They are both summaries of approaches which have been found successful in previous situations, and do not have any independent existence over and above these past applications. Therefore, they cannot be permanently *ranked:* in the moral sphere, in particular, it is impossible to judge whether a certain maxim should be considered over-riding in all conceivable circumstances. It is only *in the context of specific situations* that one may make a decision as to which of two conflicting principles should be upheld. In another situation, it would be right to come to the opposite conclusion.

In this respect, Mesthene and Rapoport are in radical opposition. Rapoport, contemplating scientific practice, defines a single over-riding principle (briefly, 'the non-coercive pursuit of truth') which can be applied universally and from which all subsidiary maxims can be deduced (or, at least, with which they are all compatible, in all instances). But Mesthene, also contemplating scientific practice, reaches the contradictory conclusion that no one principle is always over-riding or subsidiary.[11]

3. Mesthene, like Rapoport, seems to have an implicit model of the *ideal society.* But, unlike Rapoport, that society is characterized by

struggle and conflict, rather than *harmony*. There is no sense that the kind of particular, situation-bound decisions which Mesthene advocates will *lessen* the apparent conflict between the ethical principles, or lead to the universal *elimination* of any of them. Indeed, in Mesthene's world, it is only by the repeated experience of moral dilemmas, in which seemingly incompatible principles clash, that we *learn* in the ethical domain. Mesthene seems to *relish* this essential tension in the social life: he clearly does *not* consider such tension as a sign of 'instability', or a feature which is 'doomed to extinction'. And, valuing such tension in society at large, he highlights what he sees as parallel features of scientific practice. In other words, Mesthene's scientism, like Rapoport's, can be seen to be shaped by an implicit social or political model. His scientific ethic is appropriate to a differentiated and pluralist society[12], in which individuals owe allegiance to a range of social institutions and hence internalize sets of rules which conflict in many 'borderline' situations.

Questions

Which of the Mertonian norms (if any) does Mesthene draw upon?

Does Mesthene's version of a scientific ethic provide any guidance as to how one should solve the moral problems encountered in everyday life? What do you think Rapoport's ethic would recommend if it were applied to Mesthene's first example (in which an onlooker was asked by a would-be murderer which way his potential victim went)?

To what extent does Mesthene's analogy between scientific and moral principles hold good? Is it true that, when a conflict between two scientific principles (or laws, or theories) has been resolved in favor of one in a particular situation, there still remains the possibility that the other will win out in a different set of circumstances?

Assuming that you accept the analogy, what is the advantage of calling the resulting ethic a *'scientific ethic'*? Why could Mesthene not simply say: 'Ethics is like this'?

Taking the Rapoport and Mesthene essays together, do you see them as successful, logical arguments? Or would you rather describe their appeals to 'science' as being of the nature of *rhetoric*? Do you think that a successful, coherent argument of the kind that Rapoport (especially) attempts *could* be devised?

Critics of the scientific ethos

Writers (such as Rapoport) who make strong claims for a scientific ethic take for granted that the scientific community exemplifies the values chosen. However, this basic assumption can itself be challenged. Recently, a number of writers have directly criticized the Mertonian approach within the sociology of science, with its emphasis on the binding and cohesive influence of 'the ethos of science'[13]. Their main criticisms can be briefly summarized as follows.

1. The *actual behavior* of scientists is so often at variance with their professed norms that the latter cannot be so effectively binding as the Mertonians have supposed.[14] In particular, emphasis on *competition and secrecy* within the scientific community has challenged some of the more treasured images (as the debate over J. D. Watson's *The Double Helix* demonstrates).

2. The ethos cannot be homogeneous. There is evidence that it has changed over time (e.g. the eighteenth-century English amateur scientist would have little in common with a physicist at CERN); that it changes with location (e.g. as between academic and industrial scientists); and with the kind and scale of research (e.g. as between a lone, 'bench' biochemist and a team of radio astronomers).

3. The Mertonian norms are, in general, too imprecise to act as any kind of specific guide to action; and, inasmuch as they *are* meaningful, they cannot represent norms *specific* to science. Rather, they are *themselves* examples of the use of 'general cultural resources'.[15]

4. Inasmuch as scientists can be seen to demonstrate consensus, cohesion, solidarity and commitment, and the 'universal' values that transcend social divisions, these features can be seen to stem from their allegiance to 'the technical norms of paradigms, not from an overall scientific "ethos"'.[16] They are therefore inapplicable to groups which do not share these particular and specific *technical* aims: they have (in other words) a 'proper place'.

We recommend further reading in this area, by way of follow-up study. Barnes and Dolby's paper is particularly recommended.

It is important to realize that the promulgation of the image of an ideal type of scientist, exemplifying a scientific ethos, may still have an important and cohesive function within the scientific community; even if the ethos is frequently transgressed, it at least would define *the norm*. However, Barnes and Dolby consider that evidence for the existence and effectiveness of the Mertonian norms is so slender as to raise a fundamental question:

But, if evidence for the existence of these norms is so difficult to produce, how did their existence come to be postulated? The answer would appear to be that these norms have from time to time been professed by scientists... Merton can point to examples of his norms in what scientists say, but does not produce, how did their existence come to be postulated? The produce any evidence of behavior modified by these norms. Scientists typically stress such terms as rationality and scepticism in situations of celebration, justification or conflict; they are addressed to outsiders as well as to other scientists, and in the latter case ... are too nebulous to influence behavior. They are the terms of an ideology not readily convertible into behavioral recommendations; an ideology that has not always featured strongly in the scientific world...[17]

If this is so, then Merton is merely reflecting the kind of scientistic writings and utterances we have been considering. Not only is the relevance of this rhetoric to universal ethics problematic — so, too, is its relation to the real world of the scientific community.

> ### Essays, tutorial and seminar topics
>
> It is intended that the questions which we have incorporated into the text should be used as the basis for essay assignments, and for tutorial and seminar discussions.
>
> Readers are encouraged to collect their own examples of arguments similar to those of Rapoport and Mesthene, and analyze them together in seminar sessions.

Notes for Chapter 1

1. For Merton's original statement, see: Merton, R. K. (1957). 'Science and democratic social structure.' In *Social Theory and Social Structure*, pp. 604–615. Glencoe, Illinois: Free Press. For Barber's elaboration, see: Barber, B. (1972). *Science and the Social Order*. New York: Collier. For Storer's elaboration, see: Storer, N. R. (1966). *The Social System of Science*. New York: Holt, Rinehart and Winston. For a recent defence of Merton's approach, and an attack on his critics, see: Gaston, Jerry (1978). *The Reward System in British and American Science*. New York: Wiley-Interscience.
2. Rabi, I. I. (1970). *Science and the Centre of Culture*, p. 57. New York: World Publishing Co.

3. Note 2, p. 55.
4. Salomon, J. J. (1971). 'The *Internationale* of science.' *Science Studies* 1, 24. The opening pages of this essay are an elegant exposition of this theme — of which Salomon is, finally, radically critical.
5. Asimov, Isaac (1971). Guest Editorial. *Chemical and Engineering News* 49 (quoted in *New Scientist*, 20 May 1971, p. 475).
6. From a book review in *Science Forum*, December 1970.
7. Bronowski, J. (1964). *Science and Human Values*. Harmondsworth, Middx.: Penguin.
   Waddington, C. H. (1941). *The Scientific Attitude*, Ch. 8. Harmondsworth, Middx.: Penguin.
8. Bronowski, Note 7, p. 64.
9. Szent-Gÿorgi, Albert (1970). 'The mind, the brain and education.' In *Human Values and Natural Science*, pp. 48–49. Ed. by Ervin Laszlo and James B. Wilbur. New York: Gordon and Breach.
10. See David Bloor's analysis of the Kuhn/Popper debate in: Bloor, D. (1976). *Knowledge and Social Imagery*, Ch. 4. London: Routledge and Kegan Paul.
11. It is true that Mesthene might be thought to be propounding a kind of 'meta-principle' which runs like this: 'When faced with a moral dilemma, conduct a thought-experiment in which you review all the possible outcomes, in the light of the candidate principles, and choose the course of action which involves "the smallest sacrifice of other desirable goods" '. However, this is a rule of procedure, rather than an ethical principle. To Mesthene, it characterizes *the process of taking ethics seriously*. Learning about principles arises from the cumulation of experiences of taking 'serious' moral decisions in a succession of particular situations, just as the scientist's knowledge of the natural world, and of the laws and theories in which that knowledge is expressed, develops through a succession of 'serious' experiments. In advance of any particular situation (or experiment), there is no way of telling which principle (or scientific law) will 'win out'.
12. The sociologist of knowledge can make the same claim for science: namely that 'scientific knowledge' is the *form of knowledge* appropriate to a differentiated, pluralist society. See for instance: Barnes, B. (Ed.) (1972). *Sociology of Science*, Part 6. Harmondsworth, Middx.: Penguin Modern Sociology Readings.
13. For general criticisms and discussion, see:
    Barnes, S. B. and Dolby, R. G. A. (1970). 'The scientific ethos: a deviant viewpoint.' *Archiv. europ. sociol.* 11, 3–25.
    Mitroff, Ian (1974). *The Subjective Side of Science*. Amsterdam: Elsevier.
    Mulkay, Michael (1969). 'Some aspects of cultural growth in the natural sciences.' *Social Research* 36, 22–52.

Mulkay, Michael (1976). 'Norms and ideology in science.' *Social Science Information* **15**, 637–656.

Rothman, R. A. (1972). 'A dissenting view on the scientific ethos.' *Br. J. Sociol.* **23**, 102–108.

For discussions of industrial scientists, see:

Barnes, S. B. (1971). 'Making out in industrial research.' *Science Studies* **1**, 157–175.

Ellis, N. D. (1972). 'The occupation of science.' In Barnes (Ed.) *Sociology of Science*, Note 12.

Hill, Stephen C. (1974). 'Questioning the influence of a "Social System of Science": a study of Australian scientists.' *Science Studies* **4**, 135–163. The extended notes attached to this paper are a very useful introduction to, and summary of, the relevant literature.

For a discussion of scientists' 'international behavior' see:

Salomon, J. J. (1971). 'The *Internationale* of science.' *Science Studies* **1**, 23–42.

Schroeder-Gudehus, Brigitte (1973). 'Challenge to transnational loyalties: international scientific organisations after the First World War.' *Science Studies* **3**, 93–118.

14. Rothman's essay is a particularly useful summary here.
15. For a good discussion of this point, see: Barnes and Dolby, Note 13, pp. 8–14. An unexceptionable form of the ethical appeal to scientists' behavior consists in pointing to particular *exemplary lives* which embody, in a striking form, generally accepted values. In this way, for instance, Madame Curie joins the gallery of Culture Heroes, alongside explorers and missionaries, doctors, soldiers and others whose devotion to universally held values is taken to be admirable. In such cases, no *special* claim is made for science (or any other vocation): these are simply Good People.
16. Barnes and Dolby, Note 13, p. 23. Barnes and Dolby's essay demonstrates the influence of Kuhn's notion of the 'paradigm' in this discussion. Of course, if the Mertonian model can be seen as having an implicit 'political' basis, so too can the Kuhnian model. Kuhnian images of science are now as widely available as Mertonian images and are equal candidates for scientistic manipulation — often in ways, and to political purposes, that their authors never intended! See Note 10.
17. Barnes and Dolby, Note 13, pp. 12–13.

Chapter Two
Ethics and the Content of Science

> The Economy of Nature, its Checks and Balances, its Measurements of Competing Life — all This is its Great Marvel, and has an Ethic of its Own.
>
> (Henry Beston)

We turn now to a consideration of a very common form of scientistic writing — namely, writing which draws on the *results* of science to justify certain normative patterns of 'natural' (and hence 'correct') behavior.

Appeals to the 'natural' — or, perhaps more to the point, condemnations of the *'unnatural'* — are among the most powerful forms of persuasion in the repertoire of human rhetoric. Analogies between the patterns displayed in Nature[1]* and the patterns of human life and society are very frequently sensed and drawn: the metaphor of 'Laws of Nature', so central to the whole of 'Natural Science', itself consolidates the analogy between the Natural and Social worlds. Patterns perceived in Nature *suggest patterns of preferred human behavior*. A state of affairs which is 'natural' is assumed not merely to exist, but to exist for a good reason. We *excuse* an action when we say that it was a 'natural response' to a particular set of circumstances: homosexual law reform, chemical food additives, the fluoridation of water and the Women's Liberation Movement have all been *criticized* as 'going against nature'. As Mary Douglas puts it, an appeal to Nature is one of the four main 'doom points' which can be wielded by anyone wishing to impose moral constraints on his fellows.[2] Since science claims to uncover the true 'patterns of Nature', we would expect that images from scientific theory would frequently be drawn into the arena of this powerful, elemental discourse.

However, it is just as frequently claimed that this powerful tide in the affairs of men runs counter to simple principles of logic. Briefly, it is argued, no moral rules or *prescriptions* can be logically deduced from any *description* of a state of affairs — no 'ought' can be deduced from an 'is' — since the two are of *logically distinct kinds*.[3] With this view, talk of 'facts' and talk of 'values' must be sharply distinguished and kept strictly apart: moral values are hence radically *subjective*, and we *impose* them on our perceptions of the world. No factual description of the state of the world can *force* us to call it 'good' or 'evil': these evaluations are read into the description by individuals who select the

---

* Note for Chapter 2 can be found on pp. 45—50.

facts appropriately. Since appeals to 'natural patterns' are frequently made by those who wish to call them 'good' and urge us to emulate them, another way of putting this 'is/ought' distinction is to say that, *whatever* the state of affairs described, it is always in principle possible for an individual to describe an alternative state which he will consider 'better'. In logic, therefore, so it is argued, factual descriptions of Nature cannot be held to be normative.

And yet normative appeals to Nature, however logically faulty some philosophers take them to be, continue to be made, and images drawn from science are among the cultural resources deployed by those who make them. We will consider first the case of 'Evolutionary Ethics', a well documented and thoroughly criticized historical example. We will then raise questions about the contemporary scientific use of images drawn from Ecology.

Discussion point

Consider the following pairs of opposites:
  natural/supernatural
  natural/affected
  natural/artificial
  natural/adulterated

In which of these would we use the word 'natural' as a (positive) evaluation? In which is the word 'natural' used in a sense appropriate to the notion of 'natural law'?

Evolutionary ethics

Since the publication of Darwin's *Origin of Species*, many writers have sought to draw ethical and political conclusions from Darwin's description of the role of 'natural selection' in the evolution of animate Nature. This literature, and its critique, is now extensive.[4] Before proceeding further, you should sample it.

Here read: Flew, *Evolutionary Ethics,* **preferably with the excerpts from Barbour,** *Issues in Science and Religion* **as an introduction, and Quinton's or Toulmin's essay as an optional extra (see Note 4 for details).**

The intentions of evolutionary ethics are encapsulated in this text, in which Barbour summarizes C. H. Waddington's position:

> Science can provide a secure basis for ethics by discovering and exhibiting reality to be an evolutionary process tending in a certain direction, action in conformity to which is taken as right conduct.[5]

Note the appeal here to science as a source of authority and objectivity to make ethics 'secure', and to indicate 'a certain [i.e. *one*] direction'.

Unfortunately the Barbour/Waddington text remains as a kind of 'unfulfilled programme'. Very little of the writings in this area carry out this intention clearly and concisely. Perhaps the best example is to be found in Herbert Spencer's attempts to translate Darwinism into a program of political action (or, to be more accurate, of *inaction*).

Here read: Spencer, 'Society conditioned by evolution.' In Vanderpool (Ed.), *Darwin and Darwinism,* **Note 4, pp. 199–208.**

Spencer here defines an 'Is': there '*is* a real analogy between an individual and a social organism', and organisms *do* evolve and 'progress' by the ruthless action of natural selection on available variations, ensuring that only the 'fittest' survive. From this, he deduces an 'Ought': namely, that we *should not interfere* with this 'natural' process, for that process is 'good' and is a readily available and reliable guarantee of '*progress*' — 'action in conformity with which is taken as right conduct'. Darwin himself, in his later work *The Descent of Man*, expressed similar sentiments:

> We civilised men... do our utmost to check the process of elimination; we build asylums for the imbecile, the maimed, and the sick; we institute poor-laws; and our medical men exert their utmost skill to save the life of every one to the last moment... Thus, the weak members of civilised society propagate their kind. No one who has attended to the breeding of domestic animals will doubt that this must be highly injurious to the race of man.[6]

In Victorian Britain and America[7], such notions were widely available as cultural resources, and were frequently deployed, as Social Darwinism, in support of laissez-faire social policies. It was this kind of claim which provoked a bitter counter-attack, in which accusations that a 'logical error' (which G. E. Moore later called the 'naturalistic fallacy') was being made were combined with negative evaluations of Nature. T. H. Huxley's Romanes Lecture was the classic statement of this counter-position: he denied that Man was under any compulsion to make his aims and institutions congruent with the natural world:

> Cosmic evolution may teach us how the good and evil tendencies in man have come about, but, in itself, it is incompetent to

> furnish any better reason why what we call good is preferable to what we call evil than we had before... Let us understand, once and for all, that the ethical progress of society depends, not on imitating the cosmic process, still less in running away from it, but in combatting it.

Note that in the first sentence of this quotation Huxley restates the 'Is/Ought' distinction, but that the force of the second sentence relies on his implicit over-ruling of that distinction, taking Nature to be demonstrably *evil* (exemplifying that 'might is right', that 'the weakest go to the wall', and so on) and to be counteracted by human intervention.

Here read: Huxley, T. H., 'Society modified according to human standards.' In Vanderpool (Ed.), *Darwin and Darwinism,* **Note 4, pp. 209–215.**

The clash between Spencer and T. H. Huxley exemplifies the tension between those (like Spencer) who are willing to deploy Darwinian images in order to wield the 'doom point' of Nature in support of their political beliefs, and those (like Huxley) whose sympathies resist such a move, and who therefore wish to condemn it as 'illegitimate'. The logical points embedded in their argumentation are fully discussed by the authors cited. Note particularly the following features of the debate.

1. Much hinges on the ambiguity between the technical and vernacular meanings of such terms as 'progress', 'natural selection', 'the survival of the fittest', 'higher' and 'lower', and even 'evolution' itself. In everyday language, these can have strong evaluative overtones. For example, 'natural selection' can be held to imply selection according to some 'criterion of natural wisdom': Nature is supposed to be *working for the best* when she favors particular variations of organisms. Darwin himself appeared to encourage such interpretations:

> As natural selection works solely by and for the good of each being, all corporeal and mental endowments will tend to progress toward perfection.

Similarly, 'evolution' is not *just* 'change', but a change from '*lower*' to '*higher*' – where 'higher' implies 'better', by some objective standard:

> Evolution is not yet finished, organisation has not yet reached its highest level.

Again, 'fittest' – which, in evolutionary theory, means merely 'capable of survival' – can be confused with other cultural standards of 'fitness', as in the Nazi use of Darwinian language. This 'slipperiness' of the technical terms, and hence their vernacular extension 'beyond their proper domain', can account for much of the popular attraction of Darwinian images. The point is fully discussed by both Flew (especially Chapters 2 and 3) and Toulmin (especially pp. 3—15).

2. The appeal to an underlying 'law of development' can be interpreted (as Flew points out) as a claim that certain steps are *inevitable*. If this is so, then, presumably, they will come about *whatever man does:* ethics is then irrelevant! The function of such claims is then not so much that of *guidance* as of *reassurance*. Spencer was particularly apt to make such claims:

> The ultimate development of the ideal man is logically certain — as certain as any conclusion in which we place the most implicit faith; for instance that all men will die. . . Progress, therefore, is not an accident, but a necessity. Instead of civilization being artificial, it is a part of nature; all of a piece with the development of the embryo or the unfolding of a flower.[8]

Marxist writers have, traditionally, claimed an 'inevitability' of this kind for the forthcoming revolutionary triumph of the proletariat, and some have linked this with Darwinian notions. Needham's essay (Note 4) is an example. He saw 'evolutionary progress' as a matter of *increase of conscious control.* A World State in which the means of production are collectivized and run by a panel of technical experts seemed to him to be a 'higher form' of organization than a market economy in separate nation states. A transition to such a State therefore represents the next phase in the evolutionary process, and must be not only right and good but *bound to materialize* because it has 'all the authority of evolution' behind it; and yet the transition can only come about through human actions. Flew (pp. 24—27) spots a paradox here, in that the Marxist claims *both* a hard 'inevitability' *and* a crucial role for free human choice in the 'making of history'. (Spencer, in partial contrast, sees human progress as inevitable *only if particular social arrangements exist and are not interfered with.* Needham and Spencer, both drawing on Darwinian notions, reach radically different ethical and political conclusions.)

In struggling to make sense of such apparently paradoxical notions, it is difficult to avoid the conclusion that the rhetoric of 'inevitability' is not intended to *convince,* but to 'keep up one's spirits'. The function seems close to that of *religious writings.* Waddington's 'secure basis', Spencer's 'implicit faith', Julian Huxley's search for 'an intellectual prop which can support the distressed and questioning mind, and be incorporated in the common theology of the future', all support Flew's question as to. . .

> ... whether the putative direction of human evolution is being taken to be commendable as such, or only in so far as the actual direction satisfies some other standards...

and Toulmin's conclusion that...

> ... the support given by Evolution to ethics serves as a source of confidence in our moral ideas, rather than as an intellectual justification for them.

(Still less, of course, as a *criticism* of them!) In other words, these writings are *scientistic*, in the sense which we have defined: available cultural resources deriving from science are marshalled so as to add weight to certain calls to action. In this case, the 'doom point' of Nature is also mobilized: it is possible to argue, with the authority of science, that certain courses of action 'go against Nature, and your children will suffer'.

3. As was the case with the material we discussed in Chapter 1, it is possible to infer underlying social and political models within these scientific writings. This is particularly true of Spencer, whose notion of the ideal society is explicitly conflict-ridden:

> Not simply do we see that in the competition among individuals of the same kind, survival of the fittest has from the beginning furthered production of a higher type; but we see that to the unceasing warfare between species is mainly due both growth and organisation. Without universal conflict there would have been no development of the active powers.[9]

Spencer saw in Darwinian theory just the image of a conflict-ridden Nature that he needed, and so he used it to support his social model. (Needham also saw conflict, but looked forward to its inevitable elimination.) However, not everyone interpreted Darwin in the same way. Kropotkin, for instance, writing after the Spencer/Huxley debate, and supplementing Darwin's work with his own extensive field observations, took the dominant theme in Nature (and in Darwin) to be that of *co-operation within species*. He therefore drew a very different lesson:

> Happily enough, competition is not the rule either in the animal world or in mankind. It is limited among animals to exceptional periods, and natural selection finds better fields for its activity. Better conditions are created by the *elimination of competition* by means of mutual aid and mutual support...

'Don't compete! — competition is always injurious to the species, and you have plenty of resources to avoid it!' That is the *tendency* of nature, not always realized in full, but always present. That is the watchword which comes to us from the bush, the forest, the river and the ocean. 'Therefore combine — practise mutual aid! That is the surest means for giving to each and to all the greatest safety, the best guarantee of existence and progress, bodily, intellectual, moral.' That is what Nature teaches us.[10]

Many commentators (including Flew and Toulmin) have emphasized the role of previously held values in *selecting* what one takes to be the significant or dominant 'trends' or 'tendencies' or 'direction' of evolution. (Kropotkin, one notices, has to proclaim a 'tendency' which is 'not always realized in full'!). An implicit social and political model seems to be an important feature of such a selection procedure and the image of Nature so perceived itself reinforces the preferred model.[11]

In his final chapter, called 'Seeing in an evolutionary perspective', Flew offers his own interpretation of the lessons we may all properly learn from Darwinian theory. He sees a number of implications of that theory, although he does not say that their opposites are 'incompatible with' the theory — instead, he uses the much weaker (and flexible) notion of 'incongruity'. He argues that we must abandon the ideas of Special Creation, and of divinely favored moral principles; that we have to accept that all moral ideas and ideals will continue to evolve; that we should adopt a 'critical approach to all first order moral issues' and insist on 'completely naturalistic answers to questions about the nature of ethics'; and that no moral principles are 'somehow in principle beyond all need for justification and all possibility of criticism'.

One is led to ask whether Flew is not himself here selecting what little he can gather from evolutionary theory to give some 'prop' to his own liberal, humanist and atheist position — a position that he has other, independent, and *much stronger* grounds for holding. What, precisely, is gained by calling this set of beliefs an 'evolutionary perspective'?

4. We have only considered here the simple form of the debate over evolutionary ethics, as exemplified by Spencer and T. H. Huxley. It is often claimed that later writing, in which the notion of 'evolution' is extended into 'sociocultural evolution', radically changes the emphasis and counters earlier objections. A proper consideration of this issue is beyond the scope of this book. The interested reader should explore for himself the later writings of Waddington and J. S. Huxley, Fay's essay and the work of other critics.[12]

Questions and exercises

Julian Huxley writes (and Flew agrees) that:

> It makes a great difference whether we think of the history of mankind as wholly apart from the rest of life, or as a continuation of the great evolutionary process, though with special characteristics of its own.

Do you agree? Does it make a 'great difference' to ethics? If so, *spell out precisely what the difference is*.

From what you have been able to read of the literature of its proponents, do you think that evolutionary ethics has (or could have) a direct and helpful bearing on everyday moral problems? Give examples.

Do you see the emergence of Man (with consciousness, language and culture) on the earth as a process to which the notion of 'progress' or 'advance' is appropriate? If so, in what precise ways? Consider how your answer reveals your own values.

Go carefully through the Spencer extract, 'Society conditioned by evolution', noting every instance where Spencer uses an apparently *descriptive* word or phrase in an *evaluative* way. (This is a valuable exercise for every example of scientistic writing!)

In the passage quoted above, T. H. Huxley enjoins us to 'combat' the cosmic processes. What exactly do you think he means by that phrase? What attitude to Nature is implied by it? What limits are there to our freedom to carry out his advice?

Do you find Flew's notion of 'seeing in an evolutionary perspective': (*a*) helpful; and (*b*) convincing? Does Flew, in his last chapter, avoid the logical pitfalls he has diagnosed in the earlier chapters? How?

Reflect on how often you yourself use the notions 'natural' and 'unnatural' in deciding on your own actions, or in attempting to persuade others. Do you call on scientific evidence to reinforce such judgments?

Do you make a distinction between an 'evolutionary' and a 'religious' perspective? If so, what?

Ecological ethics

> An ethic, ecologically, is a limitation of freedom of action in the struggle for existence. As an ethic, philosophically, it is a differentiation of social from anti-social conduct. (Aldo Leopold)

In recent years, there has been a proliferation of attempts to mobilize the 'doom point' of Nature by ecologists and others who are influenced by their scientific writings.[13] As in evolutionary ethics, a pattern is discerned in Nature, and that pattern is given a normative role: it is as if the Barbour/Waddington text had been rewritten and that science is now providing. . .

> . . . a secure basis for ethics by discovering and exhibiting reality to be an *ecological* process, with a certain necessary pattern, action in conformity to which is taken as right conduct.

Indeed, Aldo Leopold has provided us with just this kind of concise summary:

> A thing is right when it tends to preserve the integrity, stability and beauty of the biotic community. It is wrong when it tends otherwise.[14]

The new Ethics of Nature has much in common with the old. However, there is a noticeable change in emphasis, which may be summarized by saying that whereas, a century ago, the over-riding concern was over whether Mankind would *progress*, now the concern is over whether Mankind will *survive*. Evolutionary ethics was set within a long time horizon: in ecological debate, there is doubt as to whether we will survive into the twenty-first century! In addition, there is an *urgency* in ecological rhetoric which is fresh; and it appears to have a *much larger audience*.

Almost *any* popular environmentalist paperback provides material for discussion in this book. However, clear statements of an ecological ethic are less common. We have selected the best to come to our attention to date.

Here read: Colwell, T. B., Jr. (1970). 'Some implications of the ecological revolution for the construction of value'. In E. Laszlo and J. B. Wilbur (Eds), *Human Values and Natural Science*, pp. 245–258. New York: Gordon and Breach. Consider carefully the *form of the argument* advanced by Colwell, and the extent to which that argument is vulnerable to the kind of logical objections levelled at the claims of evolutionary ethics.

Note the following features of Colwell's paper:

1. There is much in common between Colwell's argument and that of the simpler evolutionary ethics. Consider this key passage:

> What is constant to this flux is a pattern which all changes seem to follow. This pattern, a pattern which Nature as a whole exhibits (though not necessarily in all of its parts), is that of a self-recycling energy system. Natural communities that survive, find ways of returning used energy back into Nature... The whole of Nature... can be conceived of as a 'self-repairing, constructive process' which 'represents a type of equilibrium that approximates an open steady state...' Thus the 'balance' of Nature really has to do with the *way* in which natural processes relate to their environments: namely, through the efficient re-cycling of energy. When a natural community succeeds in realizing this re-cycling, it gains a relative equilibrium or 'balance' in relation to other communities. Those which do not succeed, do not survive... The concept of balance of Nature, so conceived, therefore becomes a *normative* concept for the life of natural communities. Of the many things a natural community must accomplish, it cannot fail to achieve balance in the pattern of its energy utilization. *This is the first law of the morality of Nature...* It is but a short step from talk of natural communities to our own human communities. For man and his communities are part of Nature, they are also natural. Human life takes place within the framework of the same ecological controls which govern the rest of Nature...

And Colwell talks, later on, of...

> ... redirecting the vast thrust of our civilization along more stable ecological lines.

Environmentalist rhetoric continually echoes these themes. A contrast is constantly made between 'the world in its natural balance' and 'the world in its man-made imbalance'[15], and restoration of a 'natural balance' is urged. For instance, the 'Blueprint for Survival', discussing long-term proposals to control pollution, states:

> The long-term object of these pollution control procedures is to minimise our dependence on technology as a regulator of the ecological cycles on which we depend, and *to return as much as possible to the natural mechanisms of the ecosphere,* since in all but the short-term they are much more efficient and reliable.[16]

To this kind of injunction, parallel criticisms to those levelled at the logic of evolutionary ethics can be made: *Why should* we knuckle under, and copy these alleged patterns? Can't we, as rational beings, create *new* patterns? How do you *recognize* 'the efficient recycling of energy'? (Or 'balance' — or 'imbalance'?) How 'efficient' is good *enough*? How 'well balanced' or 'stable' do we need to be? Aren't there cases where we would be justified in 'over-riding the controls'? May we not be faced with here, to paraphrase T. H. Huxley, 'not with a pattern to be copied, but a process to be combatted'?

Besides, as in the recognition of 'evolutionary trends', there is disagreement over both the facts of the case and their significance. David Lowenthal, an American ecologist, for instance, has spotted the mobilization of the 'doom point' and dislikes it:

> The notions of 'restoring the balance of nature' and 'coming to terms with nature' are... in danger of becoming shibboleths. I have collected a number of such phrases, the stock-in-trade of conservation literature: our growing mastery over the environment brings 'the revenge of an outraged nature on man'; unless we return to 'harmony', 'nature will ultimately take the matter in hand and restore equilibrium in her own drastic and remorseless way'; countries that disregard the warnings have suffered the revenge of nature, for 'nature always has the last word'; 'you cannot cheat nature out of her rights and in the long run get away with it'... In fact, the 'balance of nature' which serves as a point of departure for many of the admonitions cited does not exist. Unless one assumes that whatever is, is right, nature is full of imperfections. Many forms of life are clumsy, ineffectual and maladapted. There are more extinct than living species, and relatively few members of any species reach maturity. Man is the only exception, because man interferes with nature enough to insure the survival of most infants and children... In considering whether our impact has been creative or destructive we can *only* be anthropocentric: we are seeking our own good, not nature's; whether we deform or destroy aspects of our environment is not good or bad in itself, but only by reference to mutable human values... Man is only one of many climax-breakers which have, and may later, upset equilibria and force major biotic changes.[17]

To Colwell's 'Spencer', Lowenthal is a spirited 'T. H. Huxley'! And he has distinguished supporters: Vannevar Bush, for instance, considers that 'we must vigorously press our endeavors to upset the balance of nature in still more ways' and Theodore Dobzhansky has urged that geneticists 'prepare to take over the controls from nature if it should become necessary to correct the deficiencies of natural selection'.

No-one could accuse these two eminent American scientists of 'running away from the cosmic process'!

2. However, there are differences between the two cases, centering essentially on the emphasis in environmental rhetoric on *survival*. This appears in Colwell's paper in his discussion of what he calls 'the *ground* of human value':

> The balance of Nature provides an objective normative model which can be utilized as the *ground* of human value... It is simply a generalization of what has been observed to be a relatively constant pattern in the behavior of natural communities. Like any scientific generalization, it is subject to change. Nor does the balance of Nature serve as the source of *all* our values. It is only the *ground* of whatever other values we may develop. But these other values must be consistent with it. The balance of Nature is, in other words, a kind of ultimate value: it performs the organizational and governing function of an absolute without at the same time possessing absolute ontological status. It is a *natural* norm, not a product of human convention or supernatural authority. It says in effect to man: 'This much *at least* you must do...'

The sense in which Colwell wishes to 'read off' major, central human values *directly* from the ecological picture is very limited indeed – he explicitly allows wide variations and 'creative disagreement about specific values'. This contrasts with both the spirit and the claims of the original proponents of *evolutionary* ethics (although later writers, such as Waddington and J. S. Huxley, who emphasize sociocultural evolution, see diversity and variation of values as evolutionary goods, necessary for further advance). And the contrast is sharpened if one asks what happens if we infringe Colwell's 'first law of the morality of Nature' – if we step outside his 'boundary limits' – if we adopt values that are not 'compatible with the ecosystems of Nature'? The answer appears to be: at the worst, mankind will be physically (and quickly) eliminated from the face of the earth; at the best, we will be seriously restricted in our ability to express and realize *any values at all*. If this is so, then Colwell's injunction is not strictly an ethical matter at all – unless, that is, you consider that whether or not we should choose to survive is worth serious ethical discussion. His injunction is a necessary precondition for us to be able to afford the luxury of ethics and rational choice. And the force of Colwell's case relies very largely on *the authority and accuracy of the ecologists' predictions* and on our propensity to believe them. Such prediction is a technical matter: that *this* change in *these* variables in the ecosystem will have *those* physical results. A significant number of people appear disposed to trust and believe many of the gloomier predictions and so feel Colwell's injunction *as a real and pressing restraint:* it strikes them as having 'cash value' in a way which the more open-ended and tenuous predictions of evolutionists do not.[18] [It is, of course, true that there *are* such evolutionary predictions. The growth of medical care and social welfare, which Spencer so feared, *may* eventually lead to a serious accumulation of 'defective genes'; but this, if it occurs at all, will be some thousands of generations ahead, and the question as to whether we are under any moral obligation to those future generations is extremely problematic, and beyond the scope of this book.[19]]

One might characterize the *kind* of claims involved here by referring to three emergent 'world-views' in the history of modern science:

(a) The Newtonian world-view of physics

Although *theologies* were derived from this world-view,[20] there has been no very influential attempt to derive a secular *ethic* directly from it: so we are here dealing with *purely physical limits or restraints.*

To the girl preparing to jump off a skyscraper, we do *not* say: 'The laws of physics imply the ethical injunction that you should not jump'; rather, we say: 'The laws of physics imply that *if* you jump, you will be killed'.

This is a matter *entirely* of 'boundary conditions' for human life. We cannot 'walk on the water', however strongly our values may lead us to wish that we could.

(b) The Darwinian world-view

Although Darwinian biology does, of course, imply physical restraints and boundary limits, its main force here is that the evolutionary ethicist derives directly from it *central definitions of* 'good conduct'.

So that he might well say to the girl on the skyscraper (especially if she were particularly attractive and/or intelligent): 'The laws of evolutionary biology imply that *you should not jump,* for the sake of the values inherent in the process. You are prejudicing the progress of the human race. It's a scandalous waste of good genes... etc. But if you *do* jump, what will happen to you is a matter for the physicist.'

(c) The Ecological world-view

As Colwell spells it out, the world-view of ecology occupies a position midway between the other two: it both *suggests values* (or, at least, a 'new vision of the man—nature relationship') and *marks out boundary limits.* So, if Colwell is typical, then the ecological ethicist might say to the girl:

> We are, of course, in creative disagreement as to whether you should jump. It's entirely up to you: you have your reasons, and we can argue about it. But if you *do* decide to jump, the laws of

Nature and my ecological imperative demand that you should make arrangements with your next-of-kin for a decent burial and the efficient re-cycling of your energy content back into Nature, or you may be prejudicing the chance of future generations having any more conversations like this.

The example is not intended to be *entirely* facetious!

3. A final point returns us to the role of *implicit social and political models* in orchestrating these ecological injunctions. Ecological rhetoric (of which Colwell's paper is but one, relatively sober example) draws explicit analogies between Nature and society and constantly stresses *balance*, *stability* and *harmony* in both: the image of the 'integrated whole' is frequently evoked. It is a model of both Nature and society in which conflict is devalued:

> If nature is not a prison and earth a shoddy way-station, we must find the faith and force to affirm its metabolism as our own — or rather, our own as part of it. To do so, means nothing less than a shift in our whole frame of reference and our attitude towards life itself, a wider perception of the landscape as a creative, harmonious being where relationships of things are as real as things. Without losing our sense of a great human destiny, and without intellectual surrender, we must affirm that the world is a being, a part of our own body... [Societies] and ecosystems as well as cells have a physiology...[21]

Politics and ecology elide, as in Aldous Huxley's essay 'The Politics of Population'[22]:

> Only when we get it into our collective head that the basic problem confronting twentieth-century man is an ecological problem will our politics improve and become realistic. How does the human race propose to survive and, if possible, improve the lot and the intrinsic quality of its individual members? Do we propose to live on this planet in symbiotic harmony with our environment? Or, preferring to be wantonly stupid shall we choose to live like murderous and suicidal parasites that kill their host and so destroy themselves?

Indeed, once the 'doom point' of Nature has been effectively mobilized, radical political changes (like those proposed in 'The Blueprint for Survival') can be contemplated which otherwise would be inconceivable. As Robert Heilbroner comments:

> When the enemy is nature... rather than another social class, it is at least imaginable that adjustments [by the capitalist and

managerial classes] could be made that would be impossible in ordinary circumstances.[23]

Such is the concern for the current 'fragmentation of society'[24] that ecological apologists can advocate a return to *cultural* wholeness, integration and harmony, *because the environment is* 'whole'. For instance, Edward Goldsmith (Editor of *The Ecologist*) has written:

> It is perhaps at the scientific level that the most basic change is required. At the moment science is divided into a host of watertight compartments, each one concerned with a specialized part of the biosphere. The latter, however, is not compartmentalized in this way. It is, on the contrary, a closely integrated system that came into being over thousands of millions of years, as a simple process. By regarding its differentiated parts as separate self-sufficient fields of study, scientists, like everyone else in our society, become preoccupied with the petty and the short-term and are blind to the long-term problems that beset us.

And Max Nicholson[25] (formerly Director-General of the Nature Conservancy) has written that:

> Conservation may well prove the means of splitting the cultural atom, and of re-integrating the human culture in ways which would have been unthinkable even ten years ago.

Indeed, so close are the relationships between ecology and social models that Otis D. Duncan (1971), in his article on 'Human ecology' in *The Encyclopaedia Britannica*, can be found complaining:

> The holistic emphasis implied by the very idea of human ecology has been a continual threat to the unity of the discipline. Comprehensive treatises on the subject typically have represented expressions of social philosophy rather than empirically grounded statements of scientific theory. Indeed, numerous commentators have put forth the view that human ecology must remain primarily a philosophic viewpoint rather than aspire to the status of a systematic discipline.

The yearning for a 'whole', integrated and conflict-free society, symbolized in the image of a pure and harmonious Nature, can have many political uses. For instance, it was prominent in the 1920s in the literature of British, American and German Youth Movements, and it was developed in Fascist propaganda. In contemporary scientistic clothing, it is currently being used, by some, to justify increases in centralized control in society and, by others, to advocate the dispersal of society into relatively small and self-contained rural townships.[25] It may be that some such restraints and changes are necessary. But it is as well to ask whether cultural resources derived from science, organized by such seductive social models and harnessed to the 'doom point' of Nature, might not be releasing a tide in the affairs of men that is (by conventional definition, with its implied evaluation) 'beyond both reason and logic'. The social roots of scientism may be of more pressing concern than its logical structure.

Seminar project

Collect samples of popular environmentalist literature and analyze them, in the terms suggested in this book for:
1. their implicit social and political models;
2. their evaluative use of scientific descriptions.

There are, of course, a very large number of such works available as cheap pamphlets and paperbacks. Useful collections of extracts include:

Barr, John (Ed.) (1971). *The Environmental Handbook.* London: Ballantine/Pan.

Editors of *The Progressive* (1970). *The Crisis of Survival.* Glenview, Illinois: Scott, Foresman.

Editors of *Ramparts* (1970). *Eco-Catastrophe.* San Francisco: Canfield Press.

Irving, R. M. and Priddle, G. B. (Eds) (1971). *Crisis.* London: Macmillan.

Love, G. A. and Love, R. M. (Eds) (1970). *Ecological Crisis.* New York: Harcourt Brace Jovanovich.

Roelofs, R. T. *et al.* (Eds) (1974). *Environment and Society.* Englewood Cliffs: Prentice-Hall.

Shephard, Paul and McKinley, Daniel (Eds) (1969). *The Subversive Science.* Boston: Houghton, Mifflin.

Shephard, Paul and McKinley, Daniel (Eds) (1971). *Environ/mental.* Boston: Houghton Mifflin.

Commentaries which suggest lines of criticism include:

Neuhaus, Richard (1971). *In Defence of People.* New York and London: Collier-Macmillan.

Ridgeway, James (1971). *The Politics of Ecology.* New York: E. P. Dutton.

Weisberg, Barry (1971). *Beyond Repair.* Boston: Beacon Press.

For further study

It might be thought that we have maliciously selected 'straw men' for discussion in Chapters 1 and 2. We therefore recommend the reader to study Jacques Monod's recent book, *Chance and Necessity* (New York: Alfred Knopf, 1971; London: Collins, 1972). Monod, a Nobel Prizewinning molecular biologist, here presents a compelling 'scientific' view of Nature, and argues forcefully for an 'ethic of knowledge':

> The ethic of knowledge that created the modern world is the only ethic compatible with it, the only one capable, once understood and accepted, of guiding its evolution.

Monod's work is serious, thoughtful, influential and 'heavyweight'. Can it also be analyzed and criticized on the lines suggested in this book?

A companion volume of critical essays, John Lewis (Ed.), *Beyond Chance and Necessity* (London: Garnstone Press, 1974), is recommended as an aid in this further study. Other useful reviews of Monod's book can be found in:

The Observer (1972). 7 May.
New Scientist (1971). 9 December, 112–114.
Science (1972). 7 January, 49–50.
Technology and Culture (1972). **13**, October, 662.

See also:
Cournaud, André (1977). 'The code of the scientist and its relationship to ethics.' *Science* **198**, 699–705.

For a briefer statement of Monod's position, see his paper 'From biology to ethics' (Occasional Paper of the Salk Institute for Biological Studies, P. O. Box 1809, San Diego, California 92112).

The recent vigorous debate over 'Sociobiology' is another topic for study. For the original statement, and a popular gloss, see:
Wilson, E. O. (1975). *Sociobiology: the New Synthesis.* Cambridge, Mass.: The Belknap Press of the Harvard University Press.
Dawkins, Richard (1976). *The Selfish Gene.* London: Oxford University Press.

Exchanges between Wilson (and his supporters) and his critics (notably the Sociobiology Study Group of the [American] Science for the People organization), and other relevant discussion and comment, can be found in:
New Scientist (1976). 13 May, 342–348.
Bio Science (1976). **26,** 182–190.
New York Review of Books (1975). 13 November and 11 December.
New York Times Magazine (1975). 12 October, 38–50.
Science (1976). **191,** 1151–1155, and **192,** 736–738.
SISCON Newsletter (1977). No. 4, 3–14.
Zygon (1976). **11,** 80–154.

See also:
Barash, D. P. (1977). *Sociobiology and Behavior.* New York: Elsevier.
Dupree, A. Hunter (1977). 'Sociobiology and the natural selection of scientific disciplines.' *Minerva* **XV,** 94–101. Essay review of Wilson, *Sociobiology.*
Durrant, John (1977). 'Sociobiology: closer to myth than to science?' *Times Higher Ednl. Suppl.,* 6 May, 13.
Ebling, F. J. (1977). 'Evaluation and ethics.' *Biologist* **24,** 127–136.
Holton, Gerald (1977). 'Sociobiology: the new synthesis?' (Harvard) *Newsletter on Science, Technology and Human Values,* No. 21, 28–43.
Leach, Edmund (1977). 'Simple stereotypes.' *New Society,* 28 July, 191.
Miles, Jr., John A. (1977). 'Burhoe, Barbour, mythology and sociobiology.' *Zygon* **12,** 42–71.
Miller, Lawrence G. (1976). 'Fated genes.' *J. Hist. Behav. Sci.* **12,** 183–190. Essay review of Wilson, *Sociobiology.*
Sade, D. Stone (1975). 'The evolution of sociality.' *Science* **190,** 261–263. Review of Wilson, *Sociobiology.*
Sahlins, Marshall (1977). *The Use and Abuse of Biology.* London: Tavistock Social Science paperback.
Silcock, Bryan (1976). 'How genetic is human behaviour?' *The Sunday Times,* 6 June, 16–17.
Smith, J. Maynard (1975). 'Survival through suicide.' *New Scientist,* 28 August, 496–497. Review of Wilson, *Sociobiology.*

Notes for Chapter 2

1. For an excellent, concise historical summary of the concept of 'Nature' and its vicissitudes, see:
    Williams, Raymond (1970). 'Ideas of Nature.' *Times Lit. Suppl.*, 4 December, 1419–1421. Since reprinted in Benthall, J. (Ed.) (1972). *Ecology: the Shaping Enquiry.* London: Longman.
2. Douglas, Mary (1970). 'Environments at risk.' *Times Lit. Suppl.*, 30 September, 1273–1275. Since reprinted in Benthall (Ed.), Note 1, and also in Douglas, Mary (1975). *Implicit Meanings*, pp. 230–248. London: Routledge and Kegan Paul. This is an important and stimulating essay. The other three 'doom points' are Time, Money and God.
3. Modern discussion of this point stems from the famous remarks of David Hume: see Flew, A. G. N. (1967). *Evolutionary Ethics.* London: Macmillan Papermac, esp. Ch. 4, 'From *Is* to *Ought?*'. It is beyond the scope of this book to explore the 'Is/Ought' controversy in any depth: for further reading, consult Hudson, W. D. (Ed.) (1970). *The Is-Ought Question.* New York: St. Martin Press; and Schurr, George (1977). *Science and Ethics* (SISCON). However, it is worth pointing out here that the crucial doctrine (namely, the belief that any description or analysis of a state of affairs has no logical bearing on how that state of affairs should be valued) has *not* commanded universal acceptance amongst philosophers. In general, those philosophers centrally interested in 'the problem of knowledge' (i.e. what it is that an individual percipient can 'know') have thought that there is a valid is/ought distinction. These have tended to be English and American philosophers who (perhaps paradoxically) value science, whilst those more centrally concerned with individual or communal action (such as the Marxists or Existentialists) have not seen the distinction as valid. Clearly, people in a 'common-sense' world see facts and values as inextricably linked: to most reasonable parents, 'that is my young son with a bottle of aspirins in his hand' tends to imply 'I ought to take them away from him before he swallows any'. And it is always worth remembering, in any case, that logic is not a particularly good way of forcing anyone to do anything!
4. A selection of commentaries includes the following:
    Barbour, Ian G. (1966). *Issues in Science and Religion*, pp. 93–96, 408–414. London: SCM Press. Easily comprehended introduction to evolutionary ethics.
    Flew, *Evolutionary Ethics*, Note 3. Polemical critique of evolutionary ethical theories from a philosophical standpoint. Includes comprehensive bibliography.

Quillian, William F. (1945). *The Moral Theory of Evolutionary Naturalism*. New Haven, Conn.: Yale University Press. Detailed analysis of nineteenth-century evolutionary ethical theories.

Quinton, Anthony (1965). 'Ethics and the theory of evolution.' In *Biology and Personality*, pp. 107–131. Ed. by I. T. Ramsey. Oxford: Blackwell. Concise refutation of J. S. Huxley and C. H. Waddington, and useful discussion of the logical background to evolutionary ethics.

Fay, Charles (1970). 'Ethical naturalism and biocultural evolution.' Ed. by Ervin Lazlo and James B. Wilbur. *Human Values and Natural Science*, pp. 159–167. New York: Gordon and Breach. Counters Quinton and Flew by stressing the *cultural* elements in human evolution.

Toulmin, S. E. (1970). 'Contemporary scientific mythology.' In *Metaphysical Beliefs*, pp. 3–71. Ed. by A. MacIntyre. London: SCM Press. Includes a thorough criticism of J. S. Huxley.

Raphael, D. D. (1958). 'Darwinism and ethics.' In *A Century of Darwin*, pp. 334–359. Ed. by S. A. Barnett. London: Heinemann. More systematic analysis of J. S. Huxley and Waddington's theories than is to be found in Flew, Quinton and Toulmin.

For a selection of original sources, together with some more modern writings in the same tradition, see:

Vanderpool, Harold Y. (Ed.) (1973). *Darwin and Darwinism*. Lexington, Mass.: D. G. Heath. A very useful reader, including well-selected excerpts from Darwin, Wallace, Spencer and T. H. Huxley.

Huxley, J S. (1941). 'The uniqueness of man.' In *The Uniqueness of Man*. London: Chatto. Simple exposition of the biological standpoint underlying the author's ethical theories.

Huxley, T. H. and Huxley, J. S. (1947). *Evolution and Ethics 1893–1943*. London: Pilot Press. Includes T. H. Huxley's Romanes Lecture, 'Evolution and ethics', and J. S. Huxley's 'Evolutionary ethics' – two classic statements.

Huxley, J. S. (1953). *Evolution in Action*, Chs 5 and 6. London: Chatto. Summary of his evolutionary ethics, laying more emphasis on 'duty' and 'destiny', and less on psychoanalysis, than his 1947 version.

Needham, Joseph (1943). 'Integrative levels.' In *Time: The Refreshing River*. London: Allen and Unwin. Statement of Marxist evolutionary ethic, stressing the inevitability of progress.

Waddington, C. H. (Ed.) (1942). 'The Relations between science and ethics.' In *Science and Ethics*. London: Allen and Unwin. This article was responsible for a revival of interest in evolutionary ethics in the 1940s. The argument is condensed and rather unclear.

Waddington, C. H. (1960). *The Ethical Animal*. London: Allen and Unwin. Extended version of his 1942 article, bringing in many technical considerations from biological theory. He never actually gets round to deriving ethical conclusions from evolution!

5. This sentence occurs, in quotation marks, in Barbour, *Issues in Science and Religion*, Note 4, p. 410. Barbour cites both Waddington (1941), *The Scientific Attitude* (London: Penguin Books) and *Science and Ethics*, Note 4, but we have been unable to trace the sentence in either work. In *Science and Ethics* (p. 18), Waddington makes the claim, which he later defends, that 'the direction of evolution [is] good simply because it *is* good according to any realist definition of that concept', and a muted version of this 'strong claim' can be found in the revised edition of his *Scientific Attitude* (1948) (Penguin), p. 34. Barbour's summary is supported by Quinton, Note 4, p. 120, who refers to Waddington's 'strong theory', which 'holds that the general trend of evolution is discoverable and that the ultimate justification of moral beliefs is to be found in the extent to which they further this trend.' Whether or not Waddington wrote the sentence quoted by Barbour, readers can judge whether it fairly represents his views.

6. Darwin, Charles (1874). *The Descent of Man*, pp. 151–152. London.

7. For the British situation, see: Burrow, J. W. (1966). *Evolution and Society*. Cambridge: Cambridge University Press.
For the American situation, see: Hofstadter, Richard (1955). *Social Darwinism in American Thought*. Boston: Beacon Press.

8. Spencer, H. (1851). *Social Statics*, pp. 79–80. London: Chapman. The full passage is a superb example of scientific rhetoric, and is well worth reading.

9. Spencer, H. (1882). *The Principles of Sociology II*, pp. 240–242. New York: D. Appleton & Co.

10. Kropotkin, Peter (1902). *Mutual Aid*, pp. 74–75. London: W. Heinemann. For a good discussion on this theme, see Hofstadter, Note 7, Ch. 5, 'Evolution, ethics and society'.

11. The idea that social systems and their concomitant perceptions of Nature are mutually supportive stems from Durkheim, and is a major theme in the work of Mary Douglas, Note 2. See also Williams, Note 1.

12. See, for instance: Grene, Marjorie (1974). *The Knower and the Known*. Berkeley & London: University of California Press.
13. This is the main theme of Douglas, Note 2. See also:
    Brighton Science for People (Ecology Group) (1977). 'Images of ecology.' *Science for People* (London), No. 35, 20—21.
    Enzensberger, Hans Magnus (1974). 'A critique of political ecology.' *New Left Review*, No. 84, 3—31.
    Lowe, Philip and Worboys, Michael (1976). 'The ecology of ecology.' *Nature* **262**, 432—433.
    For a stimulating historical analysis on these lines see:
    Worster, Donald (1977). *Nature's Economy: The Roots of Ecology*. San Francisco: Sierra Club Books.
14. Leopold, A. 'An ethic of man—land relations.' Quoted in: Nash, Roderick (Ed.) (1968). *The American Environment: Readings in the History of Conservation*, p. 108. Reading, Mass: Addison-Wesley. Nash's reader is an excellent introduction to its topic. See also Nash's clear and stimulating discussion of Leopold's ethical ideas in: 'Do rocks have rights?'. *The Center Magazine*, Nov./Dec. 1977, 2—12.
15. From: Stegner, Wallace. 'The meaning of wilderness in American civilization.' Excerpt quoted in Nash (Ed.), Note 14, p. 192. Stegner argues that the wilderness should be preserved as 'a genetic reserve, a scientific yardstick' in order to allow a *scientific* criterion of 'natural balance' to be deduced.
16. *The Ecologist* (1972). **2**, January 9.
17. Lowenthal (1959/60). 'Nature and the American creed of virtue.' *Landscape* **9**, 24—25.
18. Again, the problem of *why* this should be so is addressed directly by Mary Douglas, Note 2.
19. It is, however, an important issue, and well worth exploring. See:
    Golding, Martin O. (1968). 'Ethical issues in biological engineering.' *UCLA Law Review* **15**, 443—479.
    Golding, Martin O. (1972). 'Obligations to future generations.' *The Monist* **56**, 85—99.
    Callahan, Daniel (1971). 'What obligations do we have to future generations?' *The American Ecclesiastical Review* **CLXIV**, 245—280. A response to Golding.
    Brenton Stearns, J. (1972). 'Ecology and the indefinite unborn.' *The Monist* **56**, 612—625.
    Delattre, Edwin (1972). 'Rights, responsibilities, and future persons.' *Ethics* **82**, 254—258.
    Feinberg, Joel (1974). 'The rights of animals and unborn generations.' In *Philosophy and Environmental Crisis*, pp. 43—68. Ed. by W. T. Blackstone. Athens, Georgia: University of Georgia Press.

Surber, Jere Paul (1979). 'Obligations to future generations: explorations and problemata.' *The Journal of Value Enquiry*. In press.

20. This statement requires some qualification. There have been some attempts to use Newtonian perspectives (especially as they reinforce a broader 'natural law' approach) to support generally stoic attitudes to life. Newtonian scientism can also support the conservative position that there are fixed determinate rules of conduct (with Darwinian scientism supporting the liberal position that rules of conduct are environmentally conditional and progressive). Newtonian arguments can affirm a fixed natural *and* moral order: 'As a pendulum is pulled straight to the ground, so man must walk upright (morally).' But these injunctions do not imply the sense of 'direction' of an *ethic:* rather, they have the more religious character of attempts to reinforce the *willpower*, to 'keep one at it'. (We are indebted to George Schurr for prompting this clarification.)

For an account, see Barbour, *Issues in Science and Religion*, Note 4, esp. Chs. 2 and 3.

Theological systems, of course, have moral implications, and can be used to reinforce ethical prescriptions. See Jacobs, M. C. (1971). 'The Church and the formulation of the Newtonian world view.' *Journal of European Studies* 1, 128–148. However, the *indirectness* of the relation between the scientific law and the moral code must be stressed. Thus a Newtonian Anglican Deist *might* say to the girl about to jump off the skyscraper something like this:

> If you step off, you will inevitably fall to your death by virtue of the universal force of gravitation. This force is divine in origin, since matter is inert, lifeless and incapable of moving itself. Furthermore, the harmony and order which its operation imposes on the universe is testimony to God's providential action in the world. Doubtless you were driven to considering suicide through the misfortune which resulted from the action of vile and unprincipled men with whom you were forced to have dealings. Their godless Hobbesian and Epicurean philosophies encourage their unbridled self-seeking. They fail to see that the Church of England can act as the analogue to gravitation in the social realm and impose order and structure on the relations between men. If only such men would study astronomy and apply the lessons they learn therein to their everyday dealings, they would soon see that religious adherence is not incompatible with making an honest penny or two, but that religious conformity is in everyone's interest,

both material and spiritual, just as the God-given force of gravitation ensures self-preservation, harmony and order in the material world. Think about it...

Theologians still draw freely on scientific resources to develop their conceptual systems. For a thorough survey, see Barbour's magisterial book, Note 4.
21. From Shephard, Paul (1969). In Shephard and McKinley (Eds). *The Subversive Science* (see Seminar Project list, page 42). See also Lowe and Worboys, Note 13.
22. In: Love and Love (Eds) (1970). *Ecological Crisis* (see Seminar Project list, page 42.
23. Heilbroner, Robert L. (1970). *Between Capitalism and Socialism*, pp. 283–284. New York: Random House.
24. For an example of this kind of concern, see: Bohm, David (1971). In *The Social Impact of Modern Biology*, pp. 22–35. Ed. by W. Fuller. London: BSSRS and Routledge and Kegan Paul; and Harry Rothman's pointed contribution to the ensuing discussion, pp. 36–37.
25. Nicholson, M. (1970). *The Environmental Revolution*. London: Hodder & Stoughton (also Penguin Books, 1972).
26. It is a common mistake to suppose that a particular scientific belief can *only* be used scientistically to support *one kind of* social or political model. Darwin was cited by both 'Right' and 'Left-wing' writers, and ecology is equally manipulable. For a very brief discussion of this point, using the example of 'the belief that all individuals are born equally endowed with talents and abilities', see: Barnes, Barry (1974). *Scientific Knowledge and Sociological Theory*, pp. 128–129. London: Routledge and Kegan Paul. Barnes concludes, counter-intuitively, that 'any belief may be made to serve interests; given an appropriate overall pattern of culture, any particular belief may be made to serve any particular interest.'

Chapter Three
The Social Context of Scientism

In Chapters 1 and 2, we encountered and discussed different styles of scientistic argument. The reader may feel that the distinction made in our Introduction between the 'evaluative' and 'analytic' senses of the concept of scientism has not been rigidly applied or adhered to in the reading and discussion so far, since much of the commentary has consisted of *criticism* of scientistic claims. We have been more interested in 'getting the flavor' of scientistic rhetoric and, in achieving this, evaluative considerations have not been completely banished from the reading or discussion points. For example, in Chapter 1, the factual adequacy of Merton's characterization of the norms of scientific practice did not go unchallenged; and the coherence and cogency of Rapoport's case for the adoption of a scientific ethic was also questioned. In Chapter 2, Flew's book is shown to be unambiguously polemical, being devoted to the exposure of the logical fallacies which he believes undermine any attempt to abstract ethical values from evolutionary theory. In carrying over some of his points to the sphere of ecological ethics, we encouraged the adoption of a sceptical approach to the claims of writers such as Colwell; we hinted that such ideas might not have the firm and 'objective' grounding which is often claimed, and that their widespread and unquestioned acceptance could have an irrational basis.

The analytic approach will displace the critical in this final chapter. Considerations of the validity, cogency and intellectual adequacy of scientistic thought are to be set aside in what follows; instead, we will approach scientism as a social and historical phenomenon. We should try not to think of the scientistic arguments we shall be considering as positions which may or may not command our assent but as neutral data whose appearance stands in need of explanation and understanding.[1]*

This kind of approach is comparatively novel. Almost all discussions of scientism to date have been, for many reasons, partisan. Partly, the historical connection between evolutionary ideas and racist or fascist theory has resulted in writers treating this style of argument as an intellectual perversion which should be stamped out in the interest of mental hygiene. Also, paradoxically enough, militant opposition to certain types of scientism has itself fed off certain kinds of scientistic commitment. Those who set particular store by the value-neutrality and

* Notes for Chapter 3 can be found on pp. 63–65.

intellectual integrity of science are apt to see the infiltration of science by social and political values as the pollution of a sacred domain which must be defended and purged. Only recently has it been realized that science can be treated as a cultural manifestation susceptible to social influence and determination like any other. Consequently, the systematic and scholarly investigation of the cultural impact and accompaniments of science, such as various styles of scientistic rhetoric, is only just beginning. What follows should be regarded as an introduction to a promising and as yet barely developed area of study. In particular, integration of the generality of sociological theory with the diversity of historical reality has not yet been achieved; and, as we stressed in our Introduction, the interpretation of many of our historical examples is still controversial. Nonetheless, we believe that the kind of approach that is sketched below will prove to be increasingly fruitful and, despite the initial 'distancing' of contemporary concerns from the historical raw material that is involved, may well be the source of insights that can aid our understanding of current controversy and debate. Among other things, it can help to consolidate, substantiate and elaborate the 'implicit social models' point which we have introduced in Chapters 1 and 2.

Scientism and the self-image of science

In Chapter 1 we saw how scientistic rhetoric often referred to the 'norms of science' which have been codified by Robert K. Merton and his followers. We also drew attention to the article by Barnes and Dolby which put forward the view that norms such as Rationality, Universalism and Organized Scepticism do not and cannot provide the professional scientist with criteria which he can usefully employ in choosing between alternative courses of action encountered in his day-to-day work. In explaining how Merton came to frame his account of the norms in the absence of any evidence that scientists actually used them in the course of their professional activity, Barnes and Dolby remark:

> Scientists typically stress such terms as rationality and scepticism in situations of celebration, justification and conflict... addressed to outsiders as well as to other scientists.[2]

We will begin our examination of the social determinants of scientistic rhetoric by following up this point, concentrating on the justificatory function this rhetoric plays in situations where the scientific community is faced with internal and external conflict.

Here read: King, Michael D. (1968). 'Science and the professional dilemma.' In *Penguin Social Science Survey 1968*, pp. 34–73. Ed. by J. Gould. London: Penguin Books.

Michael King's paper is the best available analysis of scientism as a response to uncertainty and dissension in the scientific community.[3] King is concerned with the way that professions react when their status and autonomy are threatened by outsiders trying to dictate how they should conduct their affairs. He suggests that such a situation can bring about an 'identity crisis', in which alternative conceptions of the distinctive skills, rights and responsibilities which constitute the profession's idea of itself are thrown up and debated. He provides an account of an example of this kind of discussion which took place in the scientific community in the 1930s, in the course of which various kinds of scientistic assertion were made by the opposing factions. A small group of scientists became dissatisfied with the relationship that existed between science and government, and formulated arguments which sought to justify the extension of the legitimate domain of scientific activity into the political arena. King argues that in order to make such a move acceptable to the broad mass of scientists, it was necessary to demonstrate that the skills which form the core of their professional image were appropriate to politics. This involved criticizing the methods that prevailed in political discussion and decision-making, and suggesting that the patterns of analysis and virtues inculcated by a scientific education and research work were superior. The passages from Sir Richard Gregory's *Nature* editorials devoted to this end are remarkably similar to examples of scientistic rhetoric quoted earlier in this book, e.g.:

> Direct contact with Nature and enquiry into her laws... produce a habit of mind which cannot be acquired in literary fields, for they are associated with a wide outlook on life more than is popularly supposed. Science is not only able to increase the comforts of life and add to material welfare but also to inspire the highest ethical thought and action. While success in science is measured solely by discovery of facts or of relationships, in politics and in public life it is gained by fluent speech and a facile pen. In scientific work, attention must be concentrated upon material fact, but the politician and the writer attach greater importance to persuasive words and phrases, and by their oratory or literary style are able to exert an influence on public affairs altogether out of proportion to their position as determined by free standards of rational value.[4]

Gregory, King suggests, was trying to reformulate the professional image of the scientific community, by showing that scientists have the right to enter politics through their possession of expertise relevant to administration and policy-making; and, furthermore, that it is their responsibility to do so since the methods currently employed therein are inappropriate and inefficient. Underlying this redefinition of the scientist's identity was the desire to increase the status, prestige and power of the profession.

Gregory's basic position was developed by J. D. Bernal into an all-embracing theory of the relationship between science and society which not only justified the scientist taking an active part in politics and administration, but indicated that he should play a dominant role in directing the development of industrial society. Bernal's Marxist theories were attacked by the conservative chemist and philosopher of science, Michael Polanyi. King sees Polanyi as representing the traditionalist response to the identity crisis — trying to preserve the established conception of the scientist's rights and responsibilities, and to maintain the autonomy of the profession by severing rather than strengthening the links with government. The interesting point as far as we are concerned is the fact that Polanyi also made use of scientistic images. Where Gregory and Bernal tried to reform the scientist's identity by showing that he was peculiarly fitted to participate in the affairs of state, Polanyi argued that the scientific community could discharge its social responsibilities by acting as a model of good government. He discerned that the professional relationships of practicing scientists constituted an exemplar of a free society.

> The world needs science today above all as an example of the good life. Spread out over the planet scientists form even today, though submerged by disaster, the body of a great and good society. Even at present scientists of Moscow and Cambridge, of Bangalore and San Francisco, respect the same standards of science: and in the depths of shattered Germany and Japan a scientist is still one of ourselves, upholding the same code of scientific work.[5]

For Gregory, the rationality of science represented the keystone on which he attempted to build a revised image of the scientist who was duty bound to enter and improve politics. The scientist's commitment to this ideal seemed to be a kind of *moral armory* which could withstand the pressures of a corrupt world. Polanyi, on the other hand, believed that the purity of science was *vulnerable* and could only be maintained by isolating science from society. This isolation was justified and emphasized by the ideals of Communism and Universalism. Science was a model community informed by ideals which could and should be copied by the outside world, but which could easily become 'tarnished' if there was too much contact between the two. This conception of the professional identity was part of an attempt to preserve the autonomy of science by reducing the area over which it had jurisdiction.

Thus, we see that scientistic rhetoric may be generated in debates about the characteristic self-image of the man of science. In attempting to reform the scientific community, apologists of conflicting positions create and redefine idealized pictures of the virtues and values inherent in the practice of science. Various types of scientism are involved to support different conceptions of the rights and responsibilities of the scientific community, conceptions which emanate from different evaluations of how the prestige and autonomy of the community can best be preserved.

Questions and exercises

In King's view, what function does the self-image of a professional community serve? Is there anything in his article which suggests that the image must correspond to the reality of professional practice?

What stresses and strains prompted Bernal and Gregory to formulate their redefinitions? Is their response best seen as a reaction to perceived threats to the prestige and autonomy of science or as an unprovoked bid to increase the power and authority of the scientific community? Or as neither?

Further exercise

Science is currently undergoing an identity crisis which, if anything, penetrates even deeper than the controversy over the social relations of science in the 1930s and 1940s. The results of the application of scientific research to weapons' development, and its condemnation as a cause of environmental degradation, have prompted a widespread questioning of the rights and responsibilities of scientific practice both inside and outside the scientific community. The deteriorating public image of science and the relationship of science to government and industrial sponsorship are recurrent themes in this debate. An interesting contribution to the discussion is Bronowski's proposal for what he terms 'the disestablishment of science',[6] not least because it can be fruitfully analyzed in the light of King's work on the

> professional dilemma and combines elements of Gregory's and Polanyi's redefinition of the professional image of science. King suggests that the identity of a profession must encompass four aspects: autonomy, expertise, commitment and responsibility. How does Bronowski's proposal engage with each of these?
>
> Would the disestablishment of science increase or decrease the autonomy of the scientific community?
>
> What new duties and tasks would be incumbent on the scientific community as a result of disestablishment, and how does Bronowski support his claim that scientists already possess the relevant expertise?
>
> How does Bronowski contrast the commitment of the scientist to that of the politician: which Mertonian norm does he invoke to make this distinction?
>
> To whom is the scientist responsible according to Bronowski, and why? How does the (supposed) existence of this responsibility favor his proposal for the disestablishment of science?

Suggested further reading

Hahn, Roger (1971). *The Anatomy of a Scientific Institution.* London and Berkeley, Calif.: University of California Press.[7] Hahn analyzes in detail the way that the professional image of the scientific community and its attendant rights and responsibilities may cohere and contrast the ideas prevalent in other sectors of society. He shows how the congruence between the aims and ideals of the French monarchy and the incipient scientific community led to the foundation of the Paris Academy of Sciences and the evolution of a set of duties and institutional practices. 'Outsider' groups formulated a contrasting image of the scientific role, which in the context of the French Revolution resulted in the dissolution of the Academy, which was unable to adapt itself to the changed political situation.

Berman, Morris (1975). ' "Hegemony" and the amateur tradition in British science.' *Journal of Social History* **8**, 30–50.

Berman, Morris (1977). *Social Change and Scientific Organisation: The Royal Institution 1799–1844.* London: Heinemann.[8] Berman offers a Marxist analysis of topics similar to Hahn's.

Scientism as social philosophy

So far we have seen how distinctively scientistic utterances come to be made by practicing scientists engaged in discussions over the social relations of their profession. We indicated that rather than taking such remarks at face value it is possible to discern underlying purposes which scientistic rhetoric may serve and reinforce. But as yet we have confined our analysis of these purposes to those which arise out of the social situation of a well established (or, in Hahn's French case or Berman's British case, an emerging) scientific community. However, we should not forget that scientism may have an audience as well as a mouthpiece. It may constitute a set of ideas that is eagerly accepted and taken up by a certain group, who do not themselves have an active professional interest in science, but who nonetheless find scientistic rhetoric peculiarly appropriate to their social situation and needs.

This sympathetic response to scientism is more likely to be found in a situation where science exists as a distinct and discernible entity on the cultural scene, but where it is not the exclusive possession of a sharply defined professional community. Where science is fully institutionalized there are rigidly maintained entrance requirements to the scientific community which tend to emphasize the differences between the man of science and the layman. But, if the boundaries between competent practitioner, dilettante and layman are blurred, the images of science that are prevalent will be most likely to be taken up by other social groups. They will not be solely specifications of the virtues and beliefs which uniquely characterize the professional and set him apart from the rest of society, but will include elements pertinent to, and drawn from, a wider spectrum of social activity. In such a situation, active participation in science need not take the form of the production of new knowledge, but may simply be a matter of attending lectures or assisting in the purchase of scientific equipment. Support for science in the preprofessional situation will be drawn from below rather than granted by some center of power. For all these reasons, when science is not professionalized but is nonetheless practiced, scientism will be not so much a professional ideology but more a general social philosophy.

This wider community of science existed in certain provincial towns in England at the time of the Industrial Revolution. Membership of such organizations as the Birmingham Lunar Society or the Manchester Literary and Philosophical Society was by no means limited to those who possessed certified scientific expertise or had published original contributions to some field of natural knowledge. Interested amateurs and laymen were permitted to join these societies and participate in their activities. Thus if we turn to the accounts that historians have furnished to explain the emergence, persistence and eventual decline of the provincial scientific societies, we should hope to gain some insights into how science and scientistic rhetoric can mesh with broader social concerns and perceptions.

Here read: Thackray, Arnold (1974). 'Natural knowledge in cultural context: the Manchester model.' *The American Historical Review* LXXIX, 672–709.[9]

Questions and exercises

What does Thackray mean when he says that the late eighteenth-century industrialists in Manchester were 'marginal men'? Do you think he gives sufficient weight to alternative explanations and do you find the evidence he presents convincing?

What sort of function did participation in a scientific society fulfil for individuals with the status of marginal men?

In what way did their perception of science as 'rational' knowledge of 'universal' laws established by a 'democratic' community attract and sustain their interest?

How did the change from the progressivist to the conservative interpretation of natural knowledge on the part of the Manchester élite show itself?

What changes in the social situation of the élite were associated with this changed perception and what does this fact indicate about the primary determinants of the image of science with which they operated?

Suggested further reading

Buck, Peter (1975). 'Order and control: the scientific method in China and the United States.' *Social Studies of Science* 5, 237–267. Suggests the scope of an analysis of this kind, and the possibilities of its generalization across different epochs, countries and cultures. Buck relates 'images of science' in twentieth-century America and China and in seventeenth-century England through the notion of a 'crisis of modernization'.

Shapin, Steven and Thackray, Arnold (1974). 'Prosopography as a research tool in the history of science: the British scientific community 1700–1900.' *History of Science* **XII**, 1–28. Explains the importance of thinking in terms of a wider community of science when considering eighteenth- and early nineteenth-century English science, and summarizes the main components of the image of science discussed in Thackray's paper.

For parallel studies of British scientific institutions, see:

Berman, M. (1972). 'The early years of the Royal Institution 1799–1810: a re-evaluation.' *Science Studies* **2**, 205–240.

Inkster, Ian (1977). 'Science and society in the metropolis: a preliminary examination of the social and institutional context of the Askesian Society of London 1796–1807.' *Annals of Science* **34**, 1–32.

Kargon, Robert H. (1977). *Science in Victorian Manchester: Enterprise and Expertise.* Manchester University Press.

Schofield, Robert E. (1963). *The Lunar Society of Birmingham: a social history of provincial science and industry in eighteenth-century England.* London: Oxford University Press.

Shapin, S. (1972). 'The Pottery Philosophical Society 1819–1835; an examination of the cultural uses of provincial science.' *Science Studies* **2**, 311–336.

Shapin, S. (1974). 'Property, patronage and the politics of science: the founding of the Royal Society of Edinburgh.' *British Journal for the History of Science* **7**, 1–41.

Shapin, S. (1974). 'The audience for science in eighteenth-century Edinburgh.' *History of Science* **12**, 95–121.

Orange, A. D. (1971). 'The British Association for the Advancement of Science: the provincial background.' *Science Studies* **1**, 315–329.

Orange, A. D. (1972). 'The origins of the British Association for the Advancement of Science.' *Br. J. Hist. Sci.* **6**, 152–176.

Orange, A. D. (1973). *Philosophers and Provincials: The Yorkshire Philosophical Society from 1822 to 1844.* York: Yorkshire Philosophical Society.

For parallel analysis of educational institutions and ideals, see:

Inkster, I. (1975). 'Science and the Mechanics' Institutes, 1820–1850: The case of Sheffield.' *Annals of Science* **32**, 451–474.

Inkster, I. (1976). 'The social context of an educational movement: a revisionist approach to the English Mechanics' Institutes, 1820–1850.' *Oxford Review of Education* **2**, 277–307.

Shapin, S. and Barnes, B. (1977). 'Science, nature and control: interpreting Mechanics' Institutes.' *Social Studies of Science* **7**, 31–74.

Similar American studies include:

Kohlstedt, Sally G. (1976). *The Formation of the American Scientific Community: The American Association for the Advancement of Science 1848—1860*. Urbana, Illinois: University of Illinois Press.

Oleson, Alexandra and Brown, Sanborn C. (Eds) (1976). *The Pursuit of Knowledge in the Early American Republic: American Scientific and Learned Societies from Colonial Times to the Civil War*. Baltimore, Md., and London: The Johns Hopkins University Press. Especially Nathan Reingold, 'Definitions and speculations: the professionalization of science in America in the nineteenth century', pp. 33—69.

Sinclair, Bruce (1974). *Philadelphia's Philosopher Mechanics: a History of the Franklin Institute 1824—1865*. Baltimore, Md., and London: The Johns Hopkins University Press.

For a related debate between historians, see:

Cantor, G. N. and Shapin, S. (1975). 'Phrenology in early nineteenth-century Edinburgh: an historiographical discussion.' *Annals of Science* 32, 195—256. (For a bibliography on phrenology, see *SISCON Newsletter* (1977), No. 4, p. 15; and Cooter, R. (1976), 'Phrenology: The provocation of progress,' *History of Science* 14, 211—234.)

Scientism and the self-understanding of society

We have studied examples of scientistic thought which build on overall characterizations of scientific activity and the knowledge it produces, but, as yet, we have not dealt with the way that particular concepts and theories can provide material for scientistic rhetoric (although the readings on phrenology, listed above, have broached the topic). While it is necessary for a group to have some kind of direct contact with science if it is to attach social significance to the *practice* of science, the *content* of science may diffuse much more widely into society. Although the precise technical details of scientific theory can only be properly understood by a very small percentage of the population, scientific concepts may gain currency with the broad mass of society, especially if the prestige and authority of science are on the increase. The changes that scientific ideas undergo when they enter into general social discourse, and the uses to which they are put, have been studied most extensively in the context of nineteenth-century America. The social historian Charles Rosenberg has played a pioneering role in this area of study, and two of his papers form the basis of this last section.

Here read: Rosenberg, Charles R. 'Scientific theories and social thought.' In *Sociology of Science*, pp. 292–305. Ed. by B. Barnes (1972). Harmondsworth, Middx.: Penguin Modern Sociology Readers.[10]

Notice the following points concerning Rosenberg's paper:

1. When theoretical concepts become part of the common vocabulary of society, they cease to be used with the rigor and precision which governs their employment by the scientific community.[11] Rosenberg suggests that this loosening of the boundaries of their application and increased fluidity should be understood in terms of metaphor and analogy. A metaphor is created when the similarity between two areas of experience is strengthened (or even created) by concepts appropriate to one sphere being applied to a second, less ordered domain. As the latter is seen through the former, the concepts transferred undergo shifts in meaning and are subtly restructured. The two levels interact, as the less well ordered area gains in coherence and intelligibility by drawing sense from the familiar and comprehensible region of experience.[12] The concepts of scientific theory represent the known, structured realm which is imposed on the more perplexing area of social reality. Rosenberg shows how ideas drawn from neurophysiology and thermodynamics underwent this process to form a fluid theory of 'nervous force', capable of ordering many different aspects of social experience. In the course of the formation of this model the meanings of the scientific concepts transposed were radically altered.

2. Not only is the meaning of the scientific concepts changed when they are taken up in social discourse, but the aims and interests which they serve is very different. Rather than being governed by an interest in the prediction and control of the natural world, they become the means of rationalizing and consolidating the social world. There are abnormal and deviant behavior to be explained, moral injunctions to be backed up by plausible sanctions, and traditional social roles to be endorsed by reference to some underlying pattern of forces which make social convention necessary and unquestionable.

3. Notice that Rosenberg uses the same intellectual strategy as Thackray to demonstrate the primacy of social needs and purposes in determining the content and employment of scientistic images and rhetoric. Thackray showed that a change in the image of science as a 'rational' endeavor and in the justificatory tasks that the image was used to perform was correlated with a change in the social situation of the Manchester elite. He used this observation to support his claim that it is the social situation of the actors which dictates how images of science are framed and used. Similarly, Rosenberg points to a change in the social attitudes and policies that were based on the theory of

heredity in the last third of the nineteenth century in America. Since there was no substantial modification of the scientific basis of the theory of heredity during this period, he concludes that it was an alteration in the social situation and perception of those who used hereditarian ideas which was responsible for the change.

Questions and exercises

How was the fluid theory of nervous force used to perform the explanatory and justificatory tasks described in point two above?

Describe the change in social policy based on the theory of heredity that occurred between 1860 and 1890.[13]

To what extent do you agree with Rosenberg's remark that modern genetic theory necessarily has conservative implications if it is accepted that behavioral characteristics are determined by heredity? Attempt to construct a justification for reformist social policies on the basis of a strong theory of inheritance.[14]

Can you think of any science-based theories/metaphors which currently inform social thought and are used in the same way as the nineteenth-century American theories of nervous force? Is Freudian theory a possible candidate? Or ecology? Or cybernetics[15] and electronic analogies in general? To what extent does the use of these theories and metaphors, or any others you can think of, trade on the prestige and authority of science?

Are there any limits on the subject matter of the scientific theories which are taken up into general social discourse? Are theories and concepts from the life sciences more likely to be metaphorically extended and used for the rationalization of social experience than ideas drawn from areas more remote from everyday concerns?

Further exercise

The use of scientific theory to bolster up and question traditional sexual roles has been explored in greater detail by

> Rosenberg and his wife. Discuss the science-based arguments that might have been marshalled by a conservative-minded, mid-nineteenth-century, middle-class American husband to persuade his wife to stay at home and attend to her domestic duties rather than join in a temperance crusade. Consider the way in which *she* might use the same science-based theory/metaphors to support *her* point of view.
>
> *Reading:* Smith-Rosenberg, Carroll and Rosenberg, Charles (1973). 'The female animal: medical and biological views of woman and her role in nineteenth-century America.' *Journal of American History* **60**, 332–356.[16]

Notes for Chapter 3

1. As we mentioned in the Introduction, our invitation here is: to suppress value judgments; to maintain a disinterested, agnostic stance towards the positions taken up and propagated in our examples; to regard scientism as a neutral, natural phenomenon; and to attempt to seek a detached understanding of the social mechanisms which support that phenomenon. All this has led to our work being itself dubbed 'scientistic' — in the *evaluative* sense. We leave readers to judge the force of this criticism and to decide whether it can be said to invalidate the whole enterprise.
2. Barnes, S. B. and Dolby, R. G. A. (1970). 'The scientific ethos: a deviant viewpoint.' *Archive of European Sociology* **11**, 13.
3. Although King's analysis is not universally accepted, we know of no published criticism of his work. But see also: Mulkay, M. (1976). 'Norms and ideologies in science.' *Social Science Information* **15**, 637–656. For a complementary analysis of the manipulation of 'images' by scientists, see: Ezrahi, Yaron (1972). 'The political resources of science.' In *Sociology of Science*, pp. 211–230. Ed. by B. Barnes. Harmondsworth, Middx.: Penguin. See also: Johnston, T. (1972). *Professions and Powers.* London: Macmillan; and Werskey, P. Gary. 'British scientists and "outsider" politics, 1931–1945.' *Science Studies* **1**, 67–83 (reprinted in Barnes (Ed.), *Sociology of Science*, pp. 231–250).
4. Quoted in King (1968), p. 53.
5. King (1968), p. 65.
6. Bronowski, Jacob (1971). 'The disestablishment of science.' *Encounter*, July, 9–16, 96 (reprinted in Fuller, W. (Ed.) (1974). *The Social Impact of Modern Biology*, pp. 233–247. London: BSSRS and Routledge and Kegan Paul). For a good discussion of

traditional scientific values under strain, see: Leitenberg, M. (1971). 'Social responsibility (II): the classical scientific ethic and strategic-weapons development.' *Impact of Science on Society*, **21**, 123—136. See also: Nelkin, D. (1978). 'Threats and promises.' *Daedalus* **107**, No. 2, 191—209.
7. A useful review of Hahn's book is: Baker, K. M. (1972). *Minerva* **X**, 502—508. See also: *Times Lit. Suppl.* (1971). 24 December, 1599; *Technology and Culture* (1972). **13**, 312; *Isis* (1972). **63**, 218; *Br. J. Hist. Sci.* (1972). **6**, 200. See also Hahn's essay in Crosland, M. P. (Ed.) (1975). *The Emergence of Science in Western Europe*, pp. 127—138. London: Macmillan. A brilliant and exciting further reading, relevant at this point, is Darnton, R. (1970). *Mesmerism and the End of the Enlightenment in France.* New York: Shocken paperback.
8. At the time of going to press, no reviews of Berman's book have been published. Historians verbally dispute his analysis, but we know of no direct challenges in print. For a parallel analysis, see: Roberts, Gerrylynn K. (1976). 'The establishment of the Royal College of Chemistry: an investigation of the social context of early-Victorian chemistry.' In *Historical Studies in the Physical Sciences* **7**, 437—485. Ed. by R. McCormmach. Princeton University Press.
9. Informal unease among historians about Thackray's analysis has only been reflected in print in one brief exchange in: *Am. hist. Rev.* (1975). **80**, 203—205. For a general discussion of the methods Thackray employs, see: Stone, Lawrence (1971). 'Prosopography.' *Daedalus* **100**, No. 1, 46—79. Robert Kargon's study of the Manchester Lit. and Phil. does not *directly* dispute Thackray's interpretation. See also: Pyenson, Lewis (1977). ' "Who the guys were": prosopography in the history of science.' *History of Science* **15**, 155—188.
10. This is a shortened version. The full paper is in: Van Tassel, D. and Hall, M. G. (Eds) (1966). *Science and Society in the United States*, pp. 137—184. Homewood, Illinois: Dorsey. Reprinted in revised form in: Rosenberg, C. (1976). *No Other Gods.* Baltimore, Md. and London: The Johns Hopkins University Press.
11. This is the point discussed at length by Toulmin in 'Contemporary scientific mythology'. In MacIntyre (Ed.) (1970). *Metaphysical Beliefs*, pp. 3—71. London: SCM Press.
12. For a fuller discussion, see: Cameron, Iain (1977). *Metaphor in Science and Society*. SISCON. A brief introduction to recent discussions, in a form relevant to an analysis of scientism, is contained in: Edge, David (1974/75). 'Technological metaphor and social control.' *New Literary History* **VI**, 135—147 (the footnotes in this paper constitute a selected bibliography).

Standard introductory references include: Black, Max (1962). *Models and Metaphors*. Ithaca, N.Y.: Cornell University Press; Schon, D. A. (1963). *Displacement of Concepts*. London: Tavistock (since republished as *Invention and the Evolution of Ideas* (1967). London: Tavistock Social Science Paperbacks); Hesse, Mary (1966). *Models and Analogies in Science*. University of Notre Dame Press. Esp. the chapter on 'The explanatory function of metaphor', pp. 157–177.

13. Further relevant material on this point can be found in Rosenberg, C. (1974). 'The bitter fruit: heredity, disease and social thought in nineteenth-century America.' *Perspectives in American History* **VIII**, 189–235 (reprinted in *No Other Gods*, Note 10, pp. 25–53). This paper is recommended for further reading.
14. See the very brief discussion in Barnes, B. (1974). *Scientific Knowledge and Sociological Theory*, pp. 128–129. London: Routledge.
15. Karl W. Deutsch is an influential exponent of the cybernetic metaphor for social thought. Particularly interesting and revealing is his early paper, 'Mechanism, organism and society: some models in natural and social science' (1951). *Philosophy of Science* **18**, 230–252. See: Barnes, Note 14, pp. 78–84. See also: Edge, D. O. (1973). 'Technological metaphor.' In *Meaning and Control*, pp. 31–59. Ed. by D. O. Edge and J. N. Wolfe. London: Tavistock. The paper in the latter volume by D. C. Bloor ('Are philosophers averse to science?', pp. 1–30) is also recommended in this context: it points to the way in which certain modern philosophers dismiss metaphor as meaningless and essentially absurd, and (following a theme which we have touched on in Chapters 1 and 2) links this position to 'implicit social models'.
16. For contrasting interpretations, see:
    Degler, Carl N. (1974). 'What ought to be and what was: women's sexuality in the nineteenth century.' *Am hist. Rev.* **79**, 1467–1490.
    Dyhouse, Carol (1976). 'Social Darwinistic ideas and the development of women's education in England, 1880–1920.' *History of Education* **5**, 41–58.
    Parsons, Gail Pat (1977). 'Equal treatment for all: American medical remedies for male sexual problems: 1850–1900.' *J. Hist. Med.* **32**, 55–71.
    Tylor, Peter L. (1977). '"Denied the power to choose the good": sexuality and mental defect in American medical practice, 1850–1920.' *J. Social Hist.* **10**, 472–489.
    See also: Hinton, Kate (1977). *Women and Science*. SISCON.

A CONTINUING TASK arising from reflection on the complete book.
Collect examples of contemporary scientific writing and consider the extent to which their production and diffusion might be 'explained' and 'understood' in terms of the 'mouthpiece-and-audience' analysis outlined in Chapter 3. Then reconsider the claims these scientific writers make... (Do not forget to include this book in the list of candidate scientific works!)

Appendix One
'Scientific Approach to Ethics'
by Anatol Rapoport

Opinions on the relation of science to ethics are usually strong ones and are often arrived at not so much by investigation of such relations as through conviction about what these relations ought to be. These opinions tend to divide those concerned with such matters into two camps. In one, the feeling runs high that science is ethically neutral, that it is concerned with what is and not with what ought to be... No such unanimity prevails in the other camp, where it is felt that connections do exist between science and ethics. This is not surprising. Those who deny that such connections exist can readily agree, for once something is declared to be nonexistent there is nothing further to be said about it. But if something is said to exist, we wish to say more about it, and the more one says, the more controversial one's opinions are likely to be...

It may help to agree, first of all, on what is meant by an ethics or an ethical system. It seems to me that in every ethics there is involved a set of choices and a set of rules governing the making of the choices, with a proviso, however, that these rules are not entirely instrumental in the pursuit of an explicit, unambiguously defined goal. This last restriction serves to differentiate an ethics from a strategy. For in a strategy, too, one has a set of choices and a set of rules for making choices, but the goal is explicit and unambiguous. Thus, the principles governing the choices of plays in a game of bridge are principles of strategy. But there is also an ethics which excludes acts defined as cheating. The 'ethics' of bridge can also be said to have a goal — for example, the assurance that the players will continue to respect one another and will continue to play — but this goal is certainly not nearly so explicit and unambiguous as the goal of winning.

In this sense, we may speak of various professional ethics as distinguished from the 'strategy' or the technique of the profession. There is an ethics in the legal and the medical professions. There is an ethics in the business community, in the military and in the underworld.

Often strategy and ethics are not easily distinguishable. For example, the saying 'honesty is the best policy' indicates that one of the ethical principles of business is seen to be also a strategic principle. On the other hand, ethics and strategy may conflict. This is dramatically shown in the frequent violations of the so-called 'rules of warfare'.

We note, next, that scientific practice also has an ethics, and, characteristically, that the ethical principles of scientific practice are intimately intertwined with strategic principles. The scientist is guided by certain rules of evidence in his definition of what is true. Furthermore, the scientist binds himself to hold and profess views (at least in regard to matters subject to scientific investigation) which he must acknowledge to be true according to those rules of evidence. Since these rules are remarkably consistent and remarkably easy to apply (compared with other rules that govern ethical decisions), the ideal of universal agreement on matters within the jurisdiction of science seems attainable in practice. Therefore, the scientist (if he is consistent) is bound to strive for universal agreement among scientists on these matters. Moreover, the agreement is to be attained neither by coercion nor by force of personal appeal but by examination of evidence alone. In other words, if 'conversion' of an opponent to one's point of view is a desideratum (such desiderata are really not included in the ethics of scientific practice but are nevertheless carried over into scientific practice from other areas), the satisfaction of such a conversion in scientific matters is complete only if the change of view comes independently of any pressure other than the weight of evidence.

Thus, even our characteristically human tendency of wishing that others thought and acted as we do becomes modified in scientific practice, because coercive measures toward these ends are pointless. Unless the conversion is made by force of evidence, it is an empty victory to achieve it.

These, then, are the ethical principles inherent in scientific practice: the conviction that there exists objective truth; that there exist rules of evidence for discovering it; that, on the basis of this objective truth, unanimity is possible and desirable; and that unanimity must be achieved by independent arrivals at convictions — that is, by examination of evidence, not through coercion, personal argument or appeal to authority.

I submit that here is a respectable chunk of an ethical system. The question before us is whether this is, characteristically, a 'professional' ethical system on a par with other such systems, like those that govern the medical, legal, military and criminal professions, or whether there is something unique about the ethics of scientific practice which makes it a particularly suitable basis for a more general system.

Before we attempt to answer this question, let us note that many professional ethical systems tend to become more than codes of conduct governing restricted professional groups. For one thing, a given culture or subculture may be ruled or dominated by some professional group, with the result that the code of that group becomes a model for an ethical system in that culture. Rather vivid examples come to mind in the case of warlike cultures and predominantly business cultures. Indeed, it appears that the ethical system prevailing in the United States is predominantly a reflection of the code of conduct prescribed for its members by the business community.

But if this be the case. . . cannot one argue that the generalization of the ethics of scientific practice to a general code of conduct, should it occur, would be merely a case of such domination, no different in principle from analogous domination by some other professional group, such as business, bureaucracy or the military?

Here we come to the fundamental differences of opinion among those who hold that there exists a connection between science and ethics. There are those who hold the view that the code of conduct of the professional scientist does have ethical implications but that such 'scientific' ethics is confined to the behavior of the scientist *qua* scientist and should not influence his behavior in his other roles. There are those who maintain that ethics derived from scientific behavior may, indeed, become the ethical basis of conduct of a whole culture but that this basis is, in principle, no different from other possible bases. And there is the extreme view (to which I subscribe) that the ethical system derived from scientific behavior is qualitatively different from other ethical systems — is, indeed, a 'superior' ethical system in a sense which I shall presently define.

This 'superiority' of scientific ethics is extremely difficult to defend, because superiority can be established only on the basis of criteria, and whatever criteria one chooses are vulnerable to the charge of being selected in order to prove the superiority which is claimed. Thus it was easy enough for Galton to claim superiority of Englishmen over African Negroes — he set up the criteria of comparison. It is equally easy for a Brahman to prove the superiority of Hindu culture over the European, and so on. Obviously, such ethnocentric traps must be avoided. We should, however, mention in passing that the very idea that one may be caught in the trap of one's own provincialism can be entertained only from the vantage point of the scientific outlook; only this outlook allows an objective comparison of several systems. Such comparison is possible only by way of analysis, and analysis is, by definition, the core of scientific inquiry. Therein lies the qualitative difference between the scientific outlook and others. The scientific outlook is the only one capable of self-examination; it is the only one that raises questions concerning its own assertions and methods of inquiry; the only one that is able to uncover provincial biases which govern our convictions and thus at least give us the opportunity to avoid such biases.

Science, like all other systems of thought, seeks answers to questions which men hold to be of importance. But whereas, in other outlooks, answers are accepted that harmonize with particular world-views peculiar to different cultural complexes, science seeks answers which are reducible to *everyone's* experience. These cannot be answers based on esoteric or mystic experience, because such experience can be common to, at most, a few. These cannot be answers based on unquestioned authority, because such authority remains unquestioned only to the extent that experiences which could lead to questioning are excluded. These cannot be answers derived from narrowly limited experience, because science puts no limits on experience. In short, the irreducible answers to scientific questions are answers linked to those irreducible experiences which can, potentially, be shared by all mankind. The situation is most marked in the physical sciences, whose results are sometimes held to be trivial by humanists, mystics, philosophers, and others, because the assertions of these sciences are about nothing but pointer readings and so do not really touch matters of profound concern to man. Trivial or not, pointer readings are unquestionably matters of universal agreement. The concepts 'larger than', 'later than', 'between', or 'three' are the same for the Norwegian and the Hottentot. There may be other bases of universal agreement, on elementary principles of kindness, beauty or the desirability of survival, but none of these supposes common denominators of human values that are completely unambiguous bases of communication or of human communality. For the scientist, the act of communication (the utilization of one nervous system by another without detriment to either) is the basic ethical act. And it is only on the most trivial level (the level of 'extensional facts' or pointer readings) that we may be sure of perfect communication. Science, then, is the only human activity which taps the really universal communality of human experience at its roots. The remarkable thing is that the tremendous edifice of knowledge which is being created on those foundations is shared by all who participate in its creation, to the same complete degree of communality.

No matter how 'Westernized' a non-Westerner may become, it is doubtful to what extent he accepts Western 'values' and rejects his own. If he rejects these completely, there is always the question of what conflicts may accompany the transformation. But the acceptance of the scientific outlook in the area of actual investigation seems to be complete, regardless of the cultural background, and there seems to be little evidence of conflict once the vantage point of scientific outlook is reached. In other words, the conversion to the scientific outlook is almost universally irreversible. It is possible for someone who believes in a magical basis of a phenomenon to reject this belief in favour of a scientific explanation, but the reverse change of view hardly ever occurs. Nor can it be argued that such unidirectional conversions result from domination by the bearer of the scientific outlook — namely, Western civilization. If that were the case, certainly conversion to Christianity (which has been energetically pursued) would have been much more universal than conversion to the scientific outlook. Not only is this not the case, but, even in the struggle within Western civilization between the religious and the scientific views on the nature of the physical and biological world, science has been unequivocally victorious.

Contrary to prevailing opinion, I should like to defend the view that science is not, like industrial capitalism, Christianity or racism, simply another culture-bound product of Western civilization, although doubtless science has been given its greatest impetus by certain developments peculiar to Western culture. Typical of ordinary components of culture is their functional interrelation. Even if we discount the extreme position of functionalist anthropology, which views every culture as a perfectly harmonious whole, still it must be admitted that, by and large, the trend towards harmony persists. Beliefs, practices and institutions tend to be accepted or rejected (unless, of course, they are imposed by force) according to whether they tend to support or to disrupt the cultural complex and its supporting world-view.

Science seems to be a notable exception. Science has had a disruptive influence on the European social order: first, it destroyed feudalism and its support, the spiritual hegemony of the established Church; now it is continuing to disrupt it by discrediting the convictions necessary for the maintenance of colonialism, of nationalism, of laissez-faire capitalism and of the authoritarian family structure. It is true that, for 300 years, science made possible the domination of the world by Western civilization. But now the trend is reversed, and it is again science which is making this domination an anachronism.

The ethics of Western culture — that is, the conventionally accepted criteria of right and wrong — are as vulnerable to the encroachment of the scientific outlook as any other ethical system. This is why it is a mistake to list science along with the other provincialisms of the West. Typically, a Westerner believes, or has believed until his beliefs were challenged by science, that Westerners should dominate the non-Westerners; that it is in the interest of national states to be more powerful than their neighbours; that private property is forever sacrosanct; that a man is not properly dressed unless he wears a noose around his neck; that the character of a child can be improved by beatings. Much of the so-called 'moral crisis' in the Western world is traceable to the undermining of some of the most tenaciously held beliefs by the in-roads of scientific views.

In view of this disruptive influence of science (disruptive in the sense of undermining holistic cultural outlooks), how are we to account for the accelerating spread of the scientific outlook? An obvious answer points to the bait of technology. Science makes technology; technology is power; men seek power and, therefore, tolerate science. I would like to argue, however, that there is another reason — namely, the ethical appeal of the scientific outlook — which marks science as not simply another of the 'white man's ways' but as something *sui generis*.

There seems to be something universally satisfying about the scientific view, at least as it affects man's outlook on his environment.

Once the vantage point of this view is attained, other views seem improverished, provincial, naive. There is no going back.

I am aware that this claim had previously been made for different religions in their expansionist phase. There may have been some justification for those claims in the sense that the acceptance of those religions (particularly Christianity, Buddhism, Islam) carried with it the exhilarating feeling of revelation. Yet there turned out to be several of these so-called 'great religions', and, in spite of the fact that they seem to have similar ethical cores, their theologies are 'incompatible', or, rather, have no basis of comparison, so that if one adheres to a great religion, one is usually either a Buddhist or a Moslem or a Jew or a Christian, or whatever, for no explainable reason, except the accident of first contacts. Science alone has succeeded in constructing a really unified philosophy, and this is because scientific philosophy is not just another philosophy — that is, another poetically harmonious system of metaphors. Scientific philosophy makes possible the examination and comparison of *philosophies*, whether as systems of logical constructs, or, more characteristically, as instances of human behavior.

The bid of scientific ethics for universal acceptance rests on the claim of science to be the first instance of a universal point of view about man's environment and, moreover, a point of view not imposed by coercion or even by power of persuasion or dramatic, personal example but by its inherent, universal appeal to universal human experience, through being rooted in reliable knowledge.

Ethics, however, is not complete unless it includes man's outlook on himself as well as on his environment. The extension from scientific outlook to scientific ethics is simply the extension of the subject matter of scientific investigation from man's environment to man himself. It is, therefore, with the consequences of this extension that we are concerned.

At this point it is proper to review the divergences in the opinions of those who hold that there is a connection between science and ethics, aside from the difference of opinion on the relevance of scientific professional ethics to general ethics. Some confine the connection to the possibility of studying various existing ethical systems by the objective methods of science. Others go further and say that ethical systems, typically, contain not only ends but also convictions or tacit assumptions concerning the most effective means to reach those ends. Admitting that ends cannot be chosen by scientific inquiry, they maintain that scientific methods are applicable to the search for effective means. For example, the elimination of conflict within a society may be an end in two different ethical systems. In one, the end may be pursued by strict apportionment of status, with attached privileges and responsibilities; in another, by an approach to an egalitarian ideal. Some would differentiate between the end and the means and maintain that the former is chosen arbitrarily, while the latter could be prescribed by a scientific investigation which could, presumably, determine the efficacy of each course of action.

Both of these views, which attribute to science only a limited role in relation to ethics, assume the possibility of sharp division: in one case between an inquiry into what *is* and a conviction about what *ought* to be; in the other, between means pursued and goals desired.

The third view — again an extreme one, which I am defending here — is that not only is science related to ethics but that science is becoming a determinant of ethics; that is, the ethics of science must become *the* ethics of humanity. I hold this view because I do not believe that one can separate either knowledge of what is from desires of what ought to be, or means from ends. To be sure, such separation can appear temporarily to be effective. The anthropologist can, for some time, describe and attain insight into a variety of ethical systems, at the same time holding on to his own. The physicist can, for a time, use scientifically ethical means (that is, pursuit of objective truth) in the service of scientifically unethical goals (for example, imposition of coercion by war). But those positions are unstable and are doomed to extinction. It is impossible, in the long run, to hold provincial views while pursuing knowledge. Comparative ethics or the dispassionate examination of means to attain arbitrarily chosen goals are not innocent pursuits. On the contrary, they make a serious impact on the investigator and on the society which he serves. They force the firing of questions aimed at the very foundations of existing ethical systems, foundations which can remain intact only if no questions are fired at them. I know of no existing culture or ethical system (as these are conventionally understood) which does not, to some degree at least, rest on a delusion. This is in no way surprising in view of the fact that every system of knowledge, including scientific knowledge, rests ultimately on some fictions. Scientific knowledge, however, is by definition that knowledge which can weather the shattering of its fictions. It is in this sense that scientific knowledge is unique. Alone among all cognitive systems, the scientific cognitive system does not shrink from the shattering of its own foundations, and when this happens, it becomes, paradoxically, more organized rather than disorganized and demoralized.

It is, therefore, possible to hope that the same 'ultra-stability' will characterize the ethical system derived from scientific practice. Certainly, this ethical system, like others, must rest on fictions, but the fictions are not sacrosanct. They can be shattered without a resulting disorganization of the system.

What are the elements of scientific ethics? Little can be said about them, because this ethic has not yet permeated human communities sufficiently deeply and so is not a result of actual practice. Certainly, the pursuit of truth 'wherever it may lead' is a paramount goal. It is tempting to suppose that a great many of these principles will be derivatives of this goal. It is tempting to suppose that a great many of these principles will coincide with those of the ethical systems of the great religions, dignity and brotherhood of man, but only as derivatives, the condition of dignity and brotherhood being most conducive to the pursuit of truth. The hope is that those activities of man which are condemned in most ethical systems but rationalized on other than ethical grounds will disappear, because their rationalizations will become untenable in the light of scientific inquiry. The same applies to quasi-ethical systems such as totalitarian ideologies and the highly specialized codes of conduct of small isolated communities. All those rest either on coercion or on exclusion of experience. Both coercion and exclusion of experience can be maintained, in the long run, only by the maintenance of sacrosanct fictions. Therefore, all coercive and provincial ethical systems depend critically on the fictions which support them. They collapse when the fictions are shattered, and their fictions are easily shattered once even the primitive elements of scientific inquiry are directed against them...

There is no sharp distinction between scientific outlook and scientific ethics. Both eschew authority — that is, coercion in any form — and probably for this reason are irresistibly attractive as means of liberating man from the bonds which, in his ignorance, fear, and ethnocentrism, he has imposed on himself.

Appendix Two
'On the Need for a Scientific Ethic'
by Emmanuel Mesthene

The use that the scientist makes of his principles is well known. In the normal course of scientific investigation, a hypothesis which explains some physical phenomenon adequately in every particular, but which runs counter to, say, the laws of inertia, cannot be held without further experimentation. Such experimentation must continue until the irreconcilability of the hypothesis with the laws is resolved. In most cases the hypothesis will fail to submit to further tests, will be declared inadequate and will give way to a new one.

Occasionally, however, a hypothesis will continue to satisfy all demands made upon it. All predictions made in its terms will be confirmed. Satisfied that his hypothesis is sound, the scientist will now turn to a re-evaluation of his laws. He does not hesitate to do this. Assuming the existence in the situation of a novel factor not explicable in terms of the current theoretic superstructure of his science, he is prepared to reformulate his assumptions so as to include the new data. He does not stand on his principles as absolute and unchanging, for to do so would be to render his problem insoluble. To consider the laws of science as inherent in nature and as such true for all time and for all situations is to negate the function of science.

Central in this procedure is recognition of the fact that the laws and principles which guide inquiry have grown, historically, from concrete situations; that they have developed out of the particular problems which have had to be solved in the continuing effort to gain reliable knowledge and, with it, increasing control of the environment. They are, in a sense, precedents. They are generalized ideas of successful solutions of past problems which it is wise to employ as guides in the analysis of similar problematic situations now occurring.

The fault of traditional morality lies in its failure to see that ethical principles are of the same *kind* as the laws and rules which guide scientists in other fields. Instead of treating their principles as the fruit of previous ethical activity ever ready to be tested and tested again in the moral situations of the present, traditional moralists arbitrarily set up a dualism, a dichotomy, and declare that although it is true enough that moral problems cannot be solved except in terms of the principles of ethics, the principles, in their turn, are in no way dependent upon the situations they are called upon to solve. In arriving at considered judgments by the use of moral principles as guides to conduct and in recognizing them as the springs of moral activity, as the causes of calculated and deliberate behavior, the moralist of this school admits the continuity of principle and problem. But he forces himself to travel down a one-way street. He fails to see that the principles themselves are arrived at by the use of the same *method* that gives him his solutions; he fails to see that continuity is a two-way, reciprocal and interacting affair.

By thus cutting off his principles from the stream of everyday thinking and acting, he deprives them of foundation and of source. This arbitrary separation, once made, severs the natural connection between principle and situation and forces the conclusion that the former is somehow innate in human nature, or that it was given to the world by some supernatural agency. Deprived, now, of the authority of experiment and test, ethical principles are graced, after the fact, with an outside, divine authority. This done, we are led to the further, necessary conclusion that the principles of morality are ultimate and unchanging, that they cannot be modified or adapted to meet changing needs, that in themselves they hold the key to every moral and social problem that can possibly develop between now and the end of time.

A holder of this view is at a loss to solve a problem that arises from a conflict of two or more of these 'ultimate' principles. A standard example is that of the man standing at a street intersection who sees another man emerge, running, from a nearby doorway and disappear around the corner. A minute later a third man, with a gun, comes out of the same doorway, runs up to our observer and asks whether his obviously intended victim turned to the right or to the left. The dilemma is clear. The impulse to truthfulness and honesty urges our man to reveal the fact that the fugitive turned left. His regard for human life and for justice as embodied in the legal code of the community tells him that he must save the victim and send the assailant on a false trail down the other street. Now honesty, truthfulness, regard for human life and respect for law are all good and praiseworthy moral principles. However, on a theory that considers them ultimate values-in-themselves, the man in our example would not know how to act. His ethical theory does not allow for a standard, more ultimate, which he can use as a basis for choice. He is forced to attempt a reconciliation of irreconcilables and as a result is unable to act.

Such predicaments are by no means rare. Almost every instance of indecision and failure to solve a problem, whether on an individual or a group basis, can be traced, in the last analysis, to a theoretical position whose guiding principles or rules of action are held to be invariable, unchangeable by definition, even in the face of the possible demands of the immediate situation.

Consider the problem of a community whose political organization is made up of three parties. Party A has the largest membership and is on the extreme right of the political horizon. Party B is numerically smaller and leftist. Both advocate some form of autocratic regime, at least for a temporary period following the forthcoming election. This makes them both bad in the estimation of Party C, which is the smallest party and whose political orientation is progressive-democratic. But Party C holds the balance of power. By not voting or by voting with the right, it can ensure a victory for Party A. On the other hand a coalition of Parties B and C can defeat the rightists. Again the dilemma is clear. The action taken by Party C will determine the outcome of the election. But on a social-ethical theory that considers all forms of authoritarianism *equally* bad without regard to the actual political situation obtaining in the community at the time of the election, Party C will be unable to take any action. In the absence of a party decision, the members of Party C will vote as individuals and the election will go to that group which will be able to draw most of these individual votes.

By failing to act as a party, Party C abdicates its function as a party. Its inability to solve the problem confronting it leads inevitably to its dissolution. The inability to act and the consequent political death are directly traceable to the theoretical position that does not permit modification and adaptation of its principles in terms of the immediate problem and of the probable consequences of each of the possible alternative solutions to the immediate problem.

The way out of the confusion of thought and action engendered by traditional ethical systems can be provided only by an ethical theory that employs in the investigation of social-ethical problems the same procedures that the scientist uses in the analysis of the problems that arise in his laboratory. Only a morality that uses its principles as instruments for inquiry into ethical situations and remembers always that they are instruments that have been forged in the fires of past experience can have the intelligent resiliency necessary to steer clear of predicaments similar to the ones described above. For only as instruments of investigation do laws, theories and principles have meaning. They are the tools with which we handle, shape and control the subject matter of inquiry, be it physical or social-ethical subject matter, and like many good tools they have to be redesigned from time to time in order to improve their efficiency.

Viewed in terms of such a scientific ethic, the 'insoluble' problems of traditional morality are meaningless. They are meaningless because they are not real problems. They arise, as we have seen, from a conflict of two or more moral principles. But we have also seen that such a conflict of principles can develop only when the principles supposedly in conflict are held to be unchanging and unchangeable. This position in turn is the result of a separation of principles from problematic situations that is historically and logically false. Moral principles develop out of the concrete situations of experience and cannot be considered apart from these situations. Therefore it is equivalent to a contradiction in terms to speak of a problem arising from a 'conflict of principles', because such a problem could not be recognized except from the standpoint of the very situations that define the principles said to be in conflict. Insoluble problems are experimentally and logically impossible. Problem and solution are two aspects of the same, continuous fact. It would be more accurate to speak of problematic-situations-to-be-solved. Such a hyphenated phrase would highlight the fact that problem and solution are no more than adjectival expressions serving to identify two aspects of the activity of inquiring. The words problem and solution derive their meaning from each other. The one cannot be understood without the other. The one cannot *exist* without the other.

These considerations imply neither complete relativity of values, nor the moral anarchy that results from such a view. They serve only to emphasize the fact that, historically and functionally, values, goods, are judgments about value, judgments that something is good or better, and that as such they and the principles which embody them are meaningful and relevant only within the ethical situations that call forth the activity of valuing. The relative importance of moral principles cannot be known except in terms of the problem requiring solution. The key to the solution of a problematic situation can be found only in the factors that constitute the situation and not in some inspiration from on high or in the intellectual acrobatics that result from the juggling of immutable principles.

Investigation into a problematic situation ends with the selection of one of the possible courses of action. A choice is made, a judgment is delivered as to which of these alternative courses solves the problem. This selection is made as a result of testing the alternatives in the imagination. The consequences of each are examined in order to determine what each implies, what principles it supports and what principles it violates, to determine which solves the problem most successfully at the smallest sacrifice of other desirable goods. There is no possibility of dilemmas such as the ones mentioned in the illustrations above because for each problem there is a solution that is better than all the other possible solutions. The fact that the one chosen may violate some existing moral principle means only that for the purposes of this particular inquiry the principle violated is less desirable than others which the solution supports. One principle is not better than another in the abstract. It is better *for* the particular situation being investigated. The criterion for judging the relative merits of moral principles is the problem to which these principles are relevant factors.

Let us return to our illustrations. The man at the street intersection will send the assailant down the wrong street and thus save the fugitive because he will consider that in this particular instance the principles of the preservation of life and of the maintenance of the legal code are more important than the principles of truthfulness and of honesty. This does not mean that the first two principles are always better than the last two, nor does it mean that truthfulness has been proved to be a false principle. Quite the contrary may be the case, given different circumstances. Consider the case of a witness at the trial of a man charged with murder. If he divulges certain information in his possession, the man will go to the electric chair. By perjuring himself, he can save the defendant's life. In this situation he judges that the principle of truthfulness is more important than that of the preservation of life, and he gives the testimony that convicts the murderer. On a theory that considers values and moral principles permanently ranked, it is impossible to account for such differences in behavior. Only a view that regards moral principles as one aspect of the total ethical situation can explain and guide moral activity. And only when so regarded can principles yield to more recent evidence when such modification is indicated by the problem under immediate investigation.

This is especially evident in our second example. Analysis of the problem with which it is faced will convince the leadership of Party C that a coalition with the leftist party would be advisable strategy for this specific situation. Instead of being forced to gaze passively at the two absolute evils 'dictatorship of the right' and 'dictatorship of the left', inquiry will reveal that the probable future consequences of such a coalition are more to the interest of the community than would be a victory for the rightist Party A. The fact that Party B is smaller than Party A means that Party C will exert greater control over the policies of the coalition than would be the case with the rightists in power. Knowing this, Party B will see to it that its temporary dictatorship is tempered by the wishes of the group by virtue of which it will win the election, and the community as a whole will be saved extremist government. Given the situation as described, many other advantages of a coalition will be brought to light as a result of observation. Thus, operating on a theory that permits observation and experiment, the third party will break the deadlock of inaction, will evaluate its principles in terms of the problem, and in acting as a party will help the community while at the same time retaining its identity and its potential for future action. Only in this way can moral principles fulfill their function, and only if thus handled can these principles permit different action given different circumstances. Varying factors would call for varying solutions. Fanatic leadership, greater strength, dirty politics — any of these or a dozen other qualities, if sufficiently strong within the makeup of the leftist party, might justify different action by Party C. But whatever the action, it would be the result of inquiry into the total situation. The appropriateness of the solution can derive only from the logic of the problem. Moral principles can have meaning only when they serve as the medium for experimental progress from problem to solution.

The statement made above that 'investigation into a problematic situation ends with the selection of one of the possible courses of action', is subject to misinterpretation. It may be maintained that the weakness of a critical theory of ethics such as the one advocated in this paper lies precisely in this necessity for ending every inquiry with an act of choice. It will be argued that selection of the proper solution from among all the possible alternatives must itself be made in terms of a criterion, and that scientific morality breaks down when it fails to allow for a final, 'most ultimate' principle on the basis of which such selection is to be made.

This argument overlooks the essential nature of the position advocated. Demonstrably, a theory which rejects the notion that principles of investigation are ultimate, cannot, in the end, appeal to an ultimate criterion for its validity. The question as to what it is in terms of which this final choice is made, can be answered only from the logic of the theory itself. The principal assumption of the present essay is that ethical and societal matters are subject to the procedures of investigation, to the method of inquiry employed in the physical sciences. Reference to these procedures shows that the scientist will choose one hypothesis from among many on the grounds that it best satisfies the facts observed. The *facts*, then, determine the choice. In physical investigations it is the facts of physics; in social-ethical investigations it is the facts of society that serve as the basis for the final act of choice... It is facts that give rise to problems, and it is facts that provide the solutions. Inquiry is the activity that links the event called problem with the event called solution. The function of ethical theory is to forge and reinforce that link, to make that activity more intelligent and more useful. But only a thorough-going scientific ethic that views itself as just one more fact among a multitude of facts can fulfill that function.